Management for Professionals

More information about this series at http://www.springer.com/series/10101

Diana Derval

Designing Luxury Brands

The Science of Pleasing Customers' Senses

Springer

Diana Derval
DervalResearch
Amsterdam, The Netherlands

ISSN 2192-8096   ISSN 2192-810X   (electronic)
Management for Professionals
ISBN 978-3-319-71555-1   ISBN 978-3-319-71557-5   (eBook)
https://doi.org/10.1007/978-3-319-71557-5

Library of Congress Control Number: 2017963865

Cover illustration: eStudio Calamar, Berlin/Figueres

Printed on acid-free paper

This Springer imprint is published by the registered company Springer International Publishing AG part of Springer Nature.
The registered company address is: Gewerbestrasse 11, 6330 Cham, Switzerland

To Rohan, my Ki & Ka

Foreword

As more and more "giants of the world" shift from developing into developed states, the understanding, accessibility, and desire for luxury goods have amassed like a snowball. In fact, luxury consumption has evolved such that the concept of luxury is no longer limited to tangible goods. In this era, storytelling, knowledge, exclusive service, and dream-realizing are sought after as the ultimate luxury. If I may be too quick to jump to a conclusion, I would say the future of the world economy depends on the successful uprising of the luxury market. However, taking part in the luxury market, I realize that this snowball does face the risk of melting. The optimism that has fallen on the luxury market may be too cheerful.

This is where Professor Derval comes in. In this era, luxury marketing is definitively one of the hottest topics. The luxury industry is in need of talents and tools to maintain its dynamism and germination. Professor Derval is not only a talent in this industry but a talent who develops tools and takes action in nurturing talents. I admire her for her vision and motivation to contribute to the luxury industry.

I myself often question the sustainability of luxury marketing, because I often fail to find the scientific explanation of many happenings of the world of luxury. This sometimes frustrates me. However, Professor Derval gives me faith in the luxury industry because she brings to the table scientific tools and approaches to creating, developing, and revamping luxury brands. I've had the pleasure to read one of her previous books "The Right Sensory Mix," and since then my understanding of the luxury market has elevated and she has brought sense to an industry driven by emotional preferences. Reading this book, I was again genuinely looking forward to acquiring more scientific tools to understand and grow the brands I service.

I hope that the profound insights contained in this work will be a source of inspiration and support to the many interested citizens around the world. Diana, I let you take it from here.

Annie Ho
General Manager Greater China
Stella McCartney, Kering Group

Reviews

Some would believe that luxury goods are all about sophistication and craftsmanship, or unabashed indulgence. However, being inspired by this book by Diana Derval, I myself was amazed by how many more little details influence luxury brands. Under Diana's deep research, luxury positioning tends to be rational, reasonable, and physiological, which has totally brought a brand new era into the rising luxury industry.

Emily Zhang, Advertising Editor, Harper's Bazaar

Author Diana tells a secret—why do people think that luxury items are the most precious masterpieces? Taking the exquisite angle of design, style, color, and other relevant persona preferences, she provides the most powerful explanation on several luxury brands' successes.

Jack Guo, Director, Innovation Management Training Center of College
of Continuing Studies, Shanghai Jiaotong University

As a fashion learner, I've already read a lot of books on fashion and luxury, but none of them are as vivid and detailed as this one, which uses many business cases, and at the same time, scientific methods to interpret the status quo and the development of luxury brands. It is an excellent book on luxury.

Lena Sun, Innovation Team, H&M

What appeals to me the most in Prof. Derval's book is the distinct viewpoint created from her years of research on human physiology and behavior, in addition to the core and fundamental theory of luxury branding, which is very impressive.

Xu Bo, Senior Relational Marketing Executive, Gucci

As a commercial trainer for a high-end home entertainment brand, I need to understand customers' lifestyle and purchasing behavior. This book from Prof. Diana Derval helped me to see what customers see, and understand the unique selling point of the brand I am working for. I need now to deliver this knowledge to our sales ambassadors!

Zoe He, Commercial Trainer, Bang & Olufsen

A unique and inspiring approach to the Luxury Marketing world. By giving readers concrete explanations, real life examples, and presenting them with renowned personalities, this book makes them feel closer to the industry, not to say entirely part of it.

Anthony Guérin, Press Public Relations, Lanvin

Create, revamp, and develop luxury brands learning from the best like Chanel and BMW. The book is packed with vivid cases and powerful tools—worth referring to.

Jadey Chen, Senior Manager, Christian Dior Couture

Diana Derval is bringing a fresh and bold qualitative perspective on luxury brand adaptation through local sensorial profiling that will make all marketers out there question their thinking process!

Jean-Baptiste Andreani, Managing Director, IFA Paris International Fashion Academy

All brands are looking for the answer on how to guarantee a sustainable development in the luxury market. Here Prof. Diana makes her appearance. First, she is elegant, perhaps because she is French, second, her experience of studying at ESSEC makes her have special views on luxury goods. In this book, she always manages to discover the secret weapon behind the marketing triumph of each brand. This book on luxury marketing doesn't instill the readers with marketing tools like other books. It provides the results of Prof. Diana's research on physiology and behavior, also some successful cases to inspire readers, so that they can unlock for themselves the mystery of the luxury market. Whether you are the elite of this industry or a freshman with interests in luxury, you will gain a lot with this book.

Tyki, International Economy and Trade student, Beijing University of Posts and Telecommunications

Diana Derval's sharp analysis and relevant case studies make total sense in the world of business and particularly luxury: understanding, shaking, amazing in order to create and grow. In one word, this book is inspiring.

Géraldine Michel, Marketing and Branding Professor, Sorbonne Business School Paris

Prof. Derval applies marketing, psychology, brand management, and surprisingly, biology. She seeks the real answers, starting from human research. Meanwhile, she also develops some tools that make me find everything fresh and new. It's a huge innovation and a major breakthrough in the industry. It's the key that can help enter the customers' hearts.

Fuli Zhang, CEO of Aiken Jeans

Professor Derval's analysis of the luxury marketing and consumer types is very sharp. This book will really help people understand what kind of person will consume what kind of luxury.

Tony Ding, Grand Club Lounge Manager, Grand Hyatt

Once again, I am very impressed with Diana Derval's new book. Through lively case studies, powerful marketing tools, and inspiring neuroscientific findings, she manages to explain the whys behind luxury mysteries. Designing luxury brands is a must read for both industry leaders and learners.

Philip Kotler, S. C. Johnson & Son Professor of International Marketing, Kellogg School of Management

Acknowledgments

This book was inspired by my dearest luxury and MBA students from IFA, ESSEC, Fudan, Sorbonne Business School, MIT, Donghua, Jiao tong, Tsinghua, GEM, IESEG, and INSEEC and by my academic colleagues, in particular Jean-Baptiste Andreani, who initiated me to fashion marketing. Efcharisto to Professor Halkias, ISM, for her guidance during my doctoral research on neuromarketing and luxury. Thank you all for your comments and ideas.

I am grateful to DervalResearch teams and shareholders, especially Srijoni, Sanjna, Sakthipriyadharshini, Gordon, Winni, Daisy, Helen, and the always fabulous and supportive Sandrine Goldie.

It was so helpful to get insights from experts and I am in debt to goldsmith Hettie Bremer, birdologist Yvonne Roelofs, retinologist Loïc Kernevez from Imagine Eyes, and pilot Hachemi Dendoune.

The field knowledge from our clients and friends was instrumental, and I say a big "merci" to all the brand executives who shared their vision and observations, and especially Annie Ho, General Manager Greater China, Stella McCartney, Kering group; Yves Bonnefont, CEO DS Automobiles; Robert-Jan Woltering, General Manager Raffles Singapore; Ahmed Gharib, General Manager Raffles Dubai; Remo Ruffini, CEO Moncler; Shawqi Ghanem, General Manager, Grand Optics; Olivier Arzel, Managing Director Asia Pacific, Christofle; Joan NG, VP Swarovski; and Grégoire Outters, General Manager Raymarine.

The book has been dramatically enhanced by FangFang's and Vlad Kolarov's illustrations and David Gardner's edits and suggestions.

I cannot thank enough my publisher Barbara Fess for her guidance, enthusiasm, and open-mindedness.

I was happy to count on the active participation of my friends and family. A special thanks to Natalie Ardet who helped with our aquatic bugs testing.

The love of my life Johan supported and petted me all the way: Danke!

Last but not least, a big thank you to you for reading this book.

Introduction

Why do some people buy a golden iPhone or Louboutin shoes? Why do some people "need" the latest LV bag or Hermès belt? What makes them spend over budget on luxury cars or cosmetics? In this book, you will find all the answers and finally understand what makes luxury brands so fascinating and successful. Throughout the chapters, we will explore the arcanes of luxury—a world packed with competition, status-seeking, envy, and an incredible drive—and visualize how our body and brain is programmed (or not) to crave luxury products. The science of pleasing customers' senses will be decoded through the analysis of 30+ case studies featuring leading but also emerging luxury brands like Tesla, Hermès, Tiffany, Louboutin, Swarovski, Jaeger-LeCoultre, Stella McCartney, and Moncler, from Paris to Dubai, and we turned all this knowledge into 6+ powerful marketing frameworks to help you design, revamp, or expand your own luxury brand, learning from the best.

The book also reveals and explains the science behind 15+ incredible facts about luxury shoppers:

1. The need for luxury is physiological
2. The main competitor of Porsche is not a car
3. Louboutin shoes are attractive because of light polarization
4. Women buy the latest luxury bags to compete with other women
5. Emirati are more into luxury than Norwegians, and it is linked to dopamine and testosterone
6. Men wear big luxury items to dissuade other men
7. Females use ornaments ranging from jewelry to surgery to show their value
8. Fundamentally, there are only three types of luxury shoppers, and we can point them out on a map
9. Wearing gold makes people more attractive as it mimicks carotenoid, a bio-marker of health
10. The sense of motion is driving the luxury industry
11. Chanel N°5 is a fragrance for competitive and powerful women, the only ones who can stand its strong chemical scent
12. Chocolate contains three addictive components making it an ideal luxury gift
13. Luxury shoppers are sensitive to magnetic fields
14. Some people can turn water (or blue dresses) into gold

15. People buy Tesla because it is made of SpaceX elements, and conquering Mars is now the ultimate way to expanding one's territory quest
16. Fashion victims love black and white with a pop of color, and it is linked to their sense of vision

In Chap. 1, we discover, through the iPhone, Tesla, and Harley Davidson cases, that in luxury also, size matters, and how hormones, male-to-male competition, and the sense of vibration have an impact on the luxury industry. We will also see how to profile and deeply connect with luxury shoppers using the Persona framework.

In Chap. 2, we investigate the neuroscience behind status-seeking, the sense of motion, and the different reactions towards luxury from one country to another, with the Moncler, Stella McCartney, and Yacht cases. We will use the Benefits framework to predict and address luxury shoppers' most hidden demands.

In Chap. 3, with the Kate Spade, Porsche, and Grand Optics cases, we dig into the physiology of female-to-female competition, explain why some luxury products are must-haves worth over budget spending, and explore the sense of colors. We will then spot the true competitors and best place to be in the luxury market using the Positioning map.

In Chap. 4, inspired by the Swarovski and Louboutin cases, we analyze the role of shiny in luxury and in the mate selection process and unveil the magnetic sense. We learn how to design unique and appealing luxury items using the Brands Codes (Fig. 1).

Fig. 1 In Luxury also, size matters (printed with DervalResearch permission)

In Chap. 5, we study the importance of expanding a luxury brand internationally and introduce the sense of time, looking at DS, Roberto Cavalli, and Shanghai Tang success stories, and grasp the importance of territory in luxury. We use the Wait Marketing 6Ms to engage luxury shoppers at the right moment, at the right place.

In Chap. 6, we dissect the sense of smell and see how brands like Chanel, Maxim's de Paris, and Montblanc involve celebrities and turn trends into classics, making the most of our biochemical urge to imitate successful individuals. We also use the Influencers' map to find the right brand ambassadors.

After reading this book, you will be able to:

- Profile and deeply connect with luxury shoppers
- Predict and address luxury customers' most hidden demands
- Spot the true competitors and the right positioning
- Design unique and appealing luxury products and experiences
- Clarify the messaging and engage luxury shoppers at the right moment
- Find the right brand ambassadors and create classics

Contents

1   **Understanding Luxury Shoppers**   1
     1.1   Introduction   1
     1.2   Who Are the Luxury Shoppers? The Golden iPhone Case   2
         1.2.1   Meet the Tuhao   2
         1.2.2   The Rise of Chinese Millionaires   2
         1.2.3   Gold and Other Preferences   4
     1.3   Hormones, Male-to-Male Competition, and Luxury   5
         1.3.1   Testosterone and Ornaments   5
         1.3.2   Dress to Impress   6
         1.3.3   Sex Ratio and Rivalry   6
     1.4   Luxury Is a Vibrating Industry   7
         1.4.1   The Sense of Vibration   7
         1.4.2   A Signature Sound: The Harley Davidson Case
               in India   11
         1.4.3   Luxury Watches and Other Complications: The Jaeger-
               Lecoultre Case   12
         1.4.4   A Silent Success: Tesla Motors   15
         1.4.5   Brand Naming: The Ermenegildo Zegna Case   15
         1.4.6   Vibrator Profiles: Business Applications   16
     1.5   Profiling Luxury Shoppers   17
         1.5.1   Profitable Customers: The American Express Case   17
         1.5.2   A Seamless Service: The Blacksocks Case   18
         1.5.3   The Personas Framework: The BMW Case   20
     1.6   Take-Aways   24
     References   25

2   **Identifying Profitable Markets**   27
     2.1   Introduction   27
     2.2   What Do Luxury Shoppers Want? The Superyacht Case   28
         2.2.1   Welcome On-board   28
         2.2.2   UHNWIs and Yachting   29
         2.2.3   Meet the Sheikh   29

2.3   Neurosciences, Status-Seeking, and Luxury Markets   30
      2.3.1   The Need for Luxury Is Physiological   30
      2.3.2   Status-Seeking and Geographical Area   31
      2.3.3   Luxury Shoppers' Drivers   32
2.4   Luxury Is All About Motion   35
      2.4.1   The Sense of Motion   35
      2.4.2   Free-Riding and White Powder   40
      2.4.3   The Rise of Moncler   41
      2.4.4   Ganbei 干杯 with Moutai   43
      2.4.5   Motion Profiles: Business Applications   44
2.5   Turning Luxury Features into Customers' Benefits   44
      2.5.1   Stella McCartney: A Designer with an Opinion   44
      2.5.2   The Benefits Framework: Dyson
              and the Contrarotator Enigma   45
2.6   Take-Aways   48
References   49

3   Finding the Right Positioning   51
    3.1   Introduction   51
    3.2   How to Design "Must-Have" Luxury Items? The Designer
          Bag Case   52
          3.2.1   The Fashion Accessories Frenzy   52
          3.2.2   From Bags to Pockets   52
          3.2.3   Which Bag Bitch Are You?   53
    3.3   The Physiology of Female-to-Female Competition   54
          3.3.1   Relational Aggression   54
          3.3.2   Becoming the Alpha Female   54
          3.3.3   Ornaments, Ranking, and Luxury   55
    3.4   Luxury Is All Black-and-White (with a Pop of Color)   56
          3.4.1   The Sense of Colors   56
          3.4.2   When Street Meets Luxury: The Y-3 Case   62
          3.4.3   Bold Colors and Affordable Luxury: The Kate
                  Spade Case   63
          3.4.4   Blending in or Standing Out? The Color Lenses Case   65
          3.4.5   Color Profiles: Business Applications   66
    3.5   Spotting the Right Positioning   66
          3.5.1   Identifying the Real Competitors: The Porsche Case   66
          3.5.2   Luxury Re-positioning: The Grand Optics Case
                  in Dubai   70
    3.6   Take-Aways   74
    References   75

4   Designing Luxury Brands 77
   4.1   Introduction 77
   4.2   How to Select a Winning Assortment? The Diamond Case 78
         4.2.1   The VIP Room 78
         4.2.2   Diamonds' 4Cs 78
         4.2.3   An Assortment Challenge 79
   4.3   The Laws of Attraction and Mate Selection 79
         4.3.1   Mate Copying 79
         4.3.2   Mr. Big or Mr. Right? 80
         4.3.3   Healthy, Wealthy, and Shiny 80
   4.4   Luxury Is Polarizing 82
         4.4.1   The Magnetic Sense 82
         4.4.2   L'Oréal Blond 91
         4.4.3   A Tiffany Blue Sky 93
         4.4.4   Silver Ever: The Christofle Case 93
         4.4.5   Polarotactic Profiles: Business Applications 95
   4.5   Designing a Unique and Recognizable Brand 95
         4.5.1   Visible Luxury: Louboutin and the "Chinese Red" 95
         4.5.2   The Brand Codes Framework: The Sofitel Experience . . . 96
         4.5.3   A Sparkle for Every Smile: The Swarovski Case 99
   4.6   Take-Aways 101
   References 101

5   Expanding Luxury Brands Internationally 105
   5.1   Introduction 105
   5.2   How to Reach Luxury Shoppers? The Cosmetics Case 106
         5.2.1   Travel Retail 106
         5.2.2   Luxury and Gifting 106
         5.2.3   Absurd Decisions 107
   5.3   Gender Polymorphism, Territory, and Luxury 107
         5.3.1   Sub-genders and Hormonal Quotient® 107
         5.3.2   Expanding One's Territory 110
         5.3.3   Affiliation, Family, and Luxury 110
   5.4   Luxury Is Timeless 111
         5.4.1   The Sense of Time 111
         5.4.2   Dining with Michelin Stars: The 10 Corso Como Case . . . 117
         5.4.3   Fashion Week: Derval Paris Haute-Couture Hats 119
         5.4.4   Selling Cars Like Luxury Shoes: The DS Case 121
         5.4.5   Importance of Being Global: The Charles Philip Case . . . 123
         5.4.6   Memory Profiles: Business Applications 125
   5.5   Reaching Luxury Shoppers with Wait Marketing 125
         5.5.1   From Milan to Bollywood: The Roberto Cavalli Case . . . 127
         5.5.2   The Wait Marketing 6Ms: The Jaguar Case 130
         5.5.3   Catch Me If You Can: The Shanghai Tang Case 132

    5.6   Take-Aways 134
    References 134

6   Building Iconic Brands 137
    6.1   Introduction 137
    6.2   How to Build an Iconic Brand? The Maison de Couture Case . . . 138
         6.2.1   Reinventing the Brand 138
         6.2.2   Launching a New Classic 138
         6.2.3   From Billboards to Instagram 139
    6.3   The Biochemistry of Imitating Successful Individuals 140
         6.3.1   Following Celebrities 140
         6.3.2   The "Chameleon Effect" 140
         6.3.3   The Habits of Successful People 141
    6.4   Luxury Has a Smell 141
         6.4.1   The Sense of Smell 141
         6.4.2   A Classic Scent: The Chanel N°5 Case 145
         6.4.3   Celebrity Fragrances: The Taylor Swift Case 146
         6.4.4   Involving Celebrities: Bienvenue chez Maxim's de Paris . . . 147
         6.4.5   Inhaler Profiles: Business Applications 148
    6.5   Finding the Right Brand Ambassadors 149
         6.5.1   Celebrity Endorsement: The Visit Dubai Case 149
         6.5.2   The Influencers' Map Framework: The Montblanc Case . . . 150
         6.5.3   Being an Opinion Leader: The PAN's Case 152
         6.5.4   Turning Trends into Classics: Fashion
                Afternoon-Tea at Raffles Dubai 154
    6.6   Take-Aways 156
    References 157

Conclusion 159

About Prof. Diana Derval 161

Books by the Same Author 163

About DervalResearch 165

Index 167

Understanding Luxury Shoppers

"I'd rather cry in a BMW car than laugh on the backseat of a bicycle!"
Ma Nuo, Model and TV Sensation (Bergstrom, 2012)

Luxury shoppers have different motivations, behaviors, and preferences, so that vaguely targeting "the rich" is not enough. Through the success stories of Tesla, Jaeger-LeCoultre, BMW, Harley-Davidson, or the golden iPhone, we will see in this chapter how to become a leading luxury brand by targeting the right customers for each product or service. The persona framework will help us decode luxury shoppers' motivation, behavior, and preferences.

1.1   Introduction

In this chapter, we get to understand luxury shoppers and investigate the following critical questions in Sect. 1.2 with the golden iPhone case:

- Who are the luxury shoppers?
- What luxury items are they looking for?
- What are they passionate about?
- Why do they spend so much money on cars or watches?
- How to target the relevant luxury shoppers?
- How to connect with mbillionaires?

We find unexpected answers exploring the science behind luxury and male-to-male competition in Sect. 1.3, and particularly the role of hormones like testosterone.

In Sect. 1.4, we analyze the sense of vibration and see how it relates to sound, as well as touch, and review its importance in luxury with the Harley Davidson, the Jaeger-LeCoultre, and the Tesla cases.

© Springer International Publishing AG, part of Springer Nature 2018
D. Derval, *Designing Luxury Brands*, Management for Professionals,
https://doi.org/10.1007/978-3-319-71557-5_1

In Sect. 1.5, we study how to profile luxury shoppers using the Persona framework with the examples of BMW and Blacksocks.

Major take-aways are shared in Sect. 1.6.

1.2   Who Are the Luxury Shoppers? The Golden iPhone Case

We were preparing a training for an Italian luxury brand when we first saw him—the tuhao. It was in a huge luxury shopping mall in the city of Nanjing, 2 h away from Shanghai by high-speed train.

1.2.1   Meet the Tuhao

We were discussing mix-and-match opportunities with the crew on the shop floor. Suddenly, he entered the shop, together with his bro's, and it was quite impressive: they were all tall, massive, and didn't look like they were joking around. The 4 male vendors decided to take a strategic tea break and we were left alone together with a small but energetic female vendor. The tuhao was aggressively mumbling on his golden iPhone "ho-ho" while browsing through the clothing racks. Our courageous vendor went straight to him but unlike what you would learn in good hospitality courses, she didn't make any eye contact with him—she just pointed at specific items and checked his reactions.

A bit of background information might be useful at this stage. The tuhao is a "new rich", a millionaire who made his fortune in industries like construction, real estate, mining, or you-don't-want-to-know-what. He is extremely busy and always on the phone doing *da mai mai* 大买卖—big business. Usually from the northern provinces of China, hence the rather massive stature, he enjoys his new lifestyle with the ladies, rich food (he grew a little *pijiudu* 啤酒肚 or wine belly), and luxury items. He is doing business and hanging around with his bro's. He can never wear too much gold or too big logos. His favorite brands are the leaders in their field: Louis Vuitton, Hermès, and Gucci (Fig. 1.1).

1.2.2   The Rise of Chinese Millionaires

China counts over 1 million millionaires—67,000 of which are identified as super-rich with a fortune of 100 million rmb (15 million USD) or more.

The super-rich are mostly private business owners but even then only a handful of them, like Jack Ma, Founder of Alibaba, are super super rich and can buy a 38 million rmb (5 million USD) helicopter without even noticing it in their budget. Many of them live in Beijing—10,000 have been located there—followed by Guangzhou, and Shanghai. Chinese millionaires are on average 39 years old, which is 15 years younger than their Western counterparts and 30% of them are women. In terms of activities, 55% of Chinese millionaires made their fortune from private businesses,

Fig. 1.1 Meet the Tuhao
(printed with DervalResearch
permission, drawing by Vlad
Kolarov)

20% from real estate, 15% from stock exchange, and 10% are high-earning salaried executives (Derval, 2016).

The proportion of millionaires by province is an accurate representation of status-seeking. People located in the east coast are on average much more into power and status, whether they were born there or moved there. We will discuss further the link between status-seeking and geographical area in Chap. 2 (Fig. 1.2).

Luxury sales are not just concentrated in Beijing or Shanghai: Hangzhou tower in the Zhejiang province generates a billion rmb (150 million USD) in sales with luxury goods and apparel. Due to strong taxes on imported luxury goods, many Chinese travel to Hong Kong or Paris and London to do their luxury shopping so that not even half of the luxury purchases are made in mainland China. The trend is even to hire the services of a "daigou", a person who will take care of buying luxury items abroad for you.

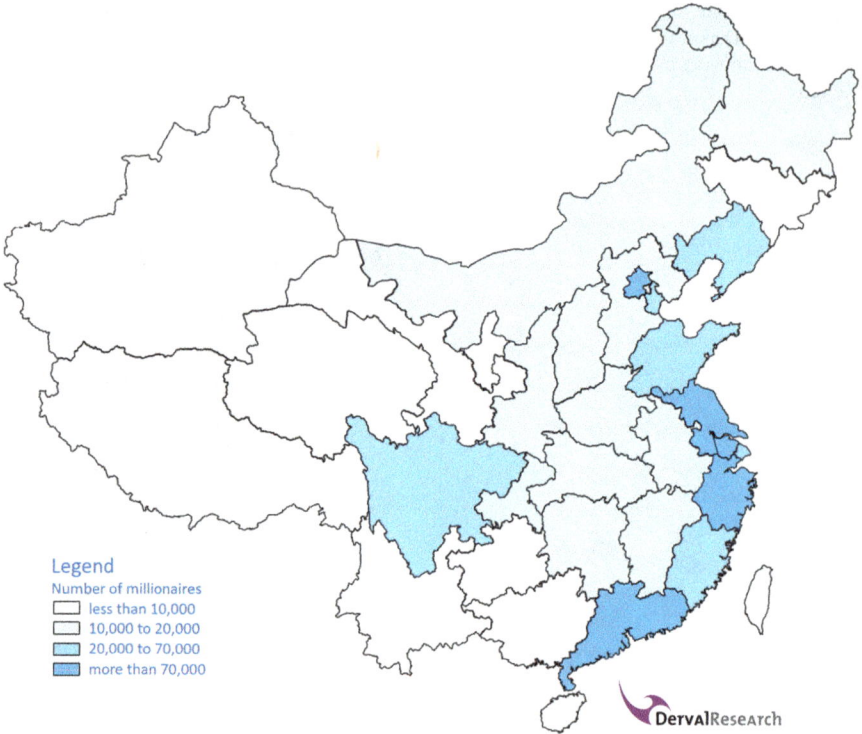

Fig. 1.2 Mapping Chinese Millionaires (printed with DervalResearch permission, source data: Hurun, 2014)

The most popular luxury items for Chinese luxury shoppers are the ones that help you make an entrance: cars, watches, fashion, and anything golden.

1.2.3   Gold and Other Preferences

The iPhone generated hundreds of billions rmb (dozens of billions USD) revenues in China where the first golden edition quickly sold out. The newest iPhone to be released recorded millions of pre-orders from China and mostly in gold and rose gold colors (Acharya, 2016).

The latest survey on Chinese millionaires' behavior and favorite brands places the iPhone as the ultimate gift even before the LV or Hermès belt, or the bottle of Moutai.

Since this millennium, China is the leading gold producer in the world, with very productive gold mines, especially in the Jiaodong peninsula (Zhang, Pian, Santosh, & Zhang, 2015). In ancient China already, powerful people would always wear gold or jade. The Emperor only was allowed to wear the color yellow. Attracted to this metal, Chinese started trading, during the "silk road", silk and other popular items against gold. No wonder that after real estate and cars, gold is extremely popular among wealthy Chinese. We will see in Chap. 4 that there is more to attraction to gold than just cultural aspects.

Now that you know more about Chinese millionaires, let's go back to our new rich in the Nanjing luxury mall. The tuhao is the type of guy with whom it is better not to make eye contact—actually if he does make eye contact it is generally a very bad sign and you are in big trouble. Luckily, he seemed to respond to the vendor's pointing at items though, still not talking but mumbling and moving his head in a manner she seemed to understand. But what is the tuhao looking for and why? The latest items? Likely. The most expensive items? Probably—as his average clothing basket would lie around 30,000 rmb (4000 USD) per shopping spree—but not *any* expensive item. For instance, suggesting that diamond encrusted pink Hello Kitty shirt might be a fatal faux-pas. More than ever, on this tense shop floor, you feel that you really need to understand your luxury shopper and suggest him the right items.

To answer this life-threatening question—you don't know the vendor's physical integrity might be at stake—we need to dig into the world of male-to-male competition.

1.3   Hormones, Male-to-Male Competition, and Luxury

In this section, we will see why and how males compete with each other and how females have their say in the whole situation.

1.3.1   Testosterone and Ornaments

Ornaments in animals play a huge role in preventing individuals from fighting, especially within the same species, as it could endanger the group: imagine if all males would get wounded and die.

In red jungle fowls, the size of the male as well as the size and chroma (saturation in color pigment) of its comb is a good predictor of male-to-male competition outcomes. The comb's appearance is directly impacted by the levels of circulating testosterone (Zuk et al., 1990). Levels of testosterone are also a good predictor of the ability to fight and, in case of defeat, to get back on our feet (or paws) and fight again. In an experiment on humans, 70% of the individuals with higher levels of testosterone would fight again, against 30% of the individuals with lower levels of testosterone. Contrary to popular belief, both men and women are exposed to both prenatal testosterone and estrogen and we will see in the following chapters that it does have an impact on how men and women perceive status (Derval & Bremer, 2012).

Accessories like jewelry, watches, or belts are, in that perspective, ornaments. No wonder they constitute a big part of the luxury expenses and generate every year over 2 billion USD revenues (Arienti, 2016). The advantage over other luxury items like a mansion or a master's painting is that you can carry these on-the-go ornaments with you. They are an extension of the human body. You might have met this director energetically pulling up his sleeves at any occasion because he is so "hands-on". Or is it maybe to showcase his gorgeous luxury watch? Or furry arms? Or often both, in an attempt to impress the team, and it often works.

If you have a chance to visit Paris museum of anthropology, the Musée du Quai Branly, you will see that the human fascination for luxury is not new and ornaments have always been at the center of preoccupations. With strong modulations between the tribes and individuals, as we will see in Chap. 2, men and women like to show and leave marks of status. In an oceanic tribe, headpieces coming in different sizes and embellishments signify the ranking and men take part in initiatic rituals. With some few exceptions like in Turkmenistan, where women tended and still tend to display their weight in gold and jewelry—especially at their wedding—it is often men in the tribe who wear ornaments.

1.3.2 Dress to Impress

The golden iPhone is in that perspective a big shiny ornament.

Back to our tuhao on the shopfloor. Who is the tuhao trying to impress: women? No, for the type of women he attracts his credit card is convincing enough. The tuhao is trying to impress other men. And it works! That is why he is not wearing a watch, but a big watch, and he is not driving a car, but a big car. Think of animals like cats or frogs trying to make themselves look bigger than they are in order to scare away competitors. Joining a meeting and exhibiting a Montblanc pen and most recently a golden iPhone will certainly impress the other participants and help in the negotiations. We will see in Chap. 6 in more details how Montblanc uses celebrities to promote its brand and revamp its business.

The overall silhouette is key, but also the vibes emitted. In horror movies, you can sense the danger by the bass sound alone. In meetings likewise, as you cannot park your Hummer in the board room, you might want to speak with a low voice to convince your peers and put on the table a phone that roars when it vibrates.

But why all this competition? Because of sex, of course, and more specifically OSR.

1.3.3 Sex Ratio and Rivalry

By OSR, understand the *Operational Sex Ratio*, or number of active males divided by the total number of active males and receptive females. The sex ratio has an influence on male behavior, and their proneness to compete or to guard their recently seduced mate (Jirotkul, 1999).

China saw, for instance, a rise in criminality due to a high sex ratio as males are pressured to own at least a house and a car in order to be eligible for marriage and some take greater risks and become criminals (Cameron, Meng, & Zhang, 2016).

To an unwealthy suitor asking her out during a Dating TV show, the model Ma Nuo wittily replied "I'd rather cry in a BMW car than laugh on the backseat of a bicycle!" and became an instant TV sensation. Although many criticized the public worshipping of money, many, also, could relate to the Beijing model's now famous quote (Bergman, 2010).

Male-to-male competition and sex ratio are driving the luxury industry, from apparel to accessories, from cars to watches, and we will see right now that it comes with vibrations.

1.4   Luxury Is a Vibrating Industry

In luxury, there is a real fascination for anything vibrating or with a motor, from watches to motorbikes. Sound and touch are vibrations and we will see in this section how they influence luxury shoppers' product preferences and purchasing behavior with the examples of Jaeger-LeCoultre, BMW, Harley-Davidson, and Tesla.

1.4.1   The Sense of Vibration

Vibrations are perceived by ears, hands, and throughout the whole body, making it a key sensory factor in the luxury decision-making process.

1.4.1.1 Sound Is a Vibration

Sound is a vibration and the number of times a sound vibrates per second is called frequency and is expressed in Hertz (Hz): a crying baby will score 6 kHz, a Harley Davidson 50 Hz and a high-end watch 1–8 Hz depending on the mechanism and finish. Sound can also be measured in terms of intensity and expressed in decibels (dB): a traditional car would score around 45 dB and a Harley Davidson could be as loud as 80 dB (Derval, 2010).

Some sounds like ultrasounds cannot be heard by humans, even though we measured fascinating variations between individuals (Fig. 1.3). During a first date, my valentine was getting distracted by a painful noise I could not hear: it was a high-pitch sound

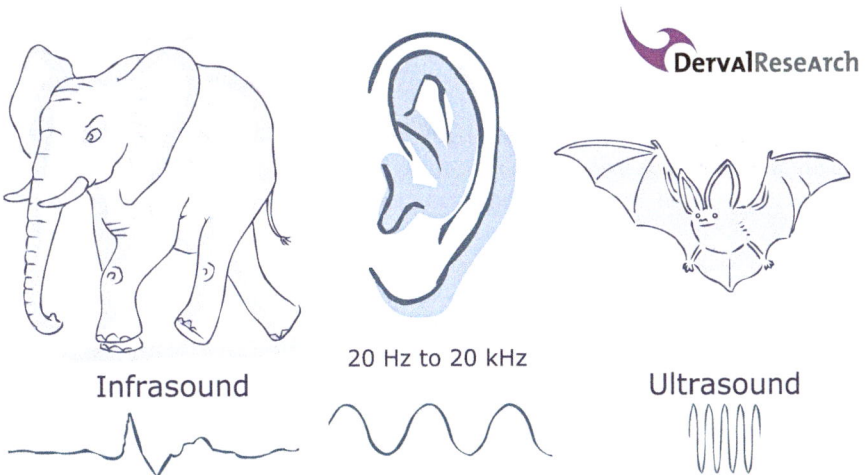

Fig. 1.3 Infrasound, sound, ultrasound (printed with DervalResearch permission)

Table 1.1 Brainwaves (Derval, 2010. Printed with DervalResearch permission)

Type	Frequency (Hz)	Activity
Delta	up to 4	Sleep
Theta	4–7	Relaxation
Alpha	8–12	Alert, working
Beta	12–30	Thinking, anxiety
Gamma	30–100+	Short term memory, objects recognition, sound, touch

Adapted from Niedermeyer and Lopes da Silva (2004)

supposed to repel mosquitos and rats, and apparently some boyfriends, too. Luckily, we threw away the guilty device and are still happy together. Similarly, infrasounds cannot be heard by humans, but some agree that the low vibrations emitted by the tuhao entering the shop could be sensed. Whales, elephants, rhinos, and tigers can produce sounds below 20 Hz: these infrasounds can cut through buildings, deep forests, and even mountains. A tiger roaring around 18 Hz will even paralyze its prey (American Institute of Physics 2000). Therefore, infrasounds are now being used as sonic weapons.

Even when sleeping, we emit vibrations or brainwaves (Table 1.1). Relaxed, we emit vibrations between 1 and 7 Hz and sound a bit like a Rolex. Sounds coming from an engine can be qualified in terms of noise, vibration, and harshness. An engine is assessed based on the number of revolutions per minute or rpm with 1000–5000 rpm for an average car versus 18,000 rpm for a Formula 1 race car (Farley, 2014). If you would play a racing F1 car on your guitar—no, but some did—a V6 engine would have a noise at a frequency of 600 Hz, and a V8 engine at a frequency of 375 Hz, which is definitely a higher pitch than a Harley Davidson (Math, 2013).

Horses are special mammals able to emit two sound tracks at the same time and it is not completely understood yet how they accomplish this vocal prowess. One track indicates the valence—positive or negative emotion—and the other highlights the intensity of the emotion. So that you know, a happy horse would typically produce shorter whinnies with a lower maximum frequency whereas an unhappy horse would produce longer whinnies in a higher frequency (Prigg, 2015).

1.4.1.2 Touch Is a Vibration

Touch is a vibration. Similarly to ears leading sounds to haircells located in the inner-ear, turning sound into electric impulses, fingerprints lead vibrations to touch receptors located in hands and skin. In addition to receptors specialized in heat and separate ones helping to perceive cold, there are four types of touch and vibration receptors:

- Light-touch receptors sensitive to mosquitos and tickling (Fig. 1.4, #1 and #2)
- Strong-touch and pressure receptors involved in grip and holding objects (#5)
- Very-low vibrations receptors responding to 50 Hz sounds like roaring tigers and Harley Davidson (#4)
- Low-vibration receptors responding to 200–300 Hz sounds like a V8 engine (#6)

This is why we do not just hear sounds but we feel their vibration throughout the whole body. Going back to the golden iPhone example, some people have been

Skin Sensors

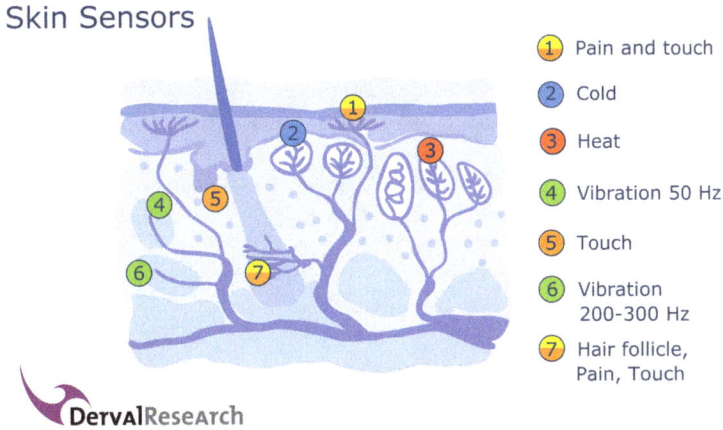

1. Pain and touch
2. Cold
3. Heat
4. Vibration 50 Hz
5. Touch
6. Vibration 200-300 Hz
7. Hair follicle, Pain, Touch

DervalResearch

Fig. 1.4 Touch receptors (Derval, 2010. Printed with DervalResearch permission)

complaining about the iPhone 6 vibration not being intense enough. CEOs of companies even switched back to the previous iPhone because of that. They were definitely missing the roaring effect during their board meetings. Even with the iPhone 7, the vibration can be sensed in the pocket but not heard, which fuels much discussion in social media. It is possible to create a personalized iPhone vibration pattern but not to alter the frequency: a major miss for a luxury ornament, supposed to help impress (or paralyze?) counterparts and teams during business meetings.

1.4.1.3 Vibrator Profiles

We can identify three types of luxury shoppers based on their perception of vibrations (Fig. 1.5):

– Non-vibrators. They do not hear bass so well and can therefore enjoy a ride on a Harley-Davidson. In terms of touch, they do not perceive fine textures and prefer slick surfaces like an iPhone
– Medium-vibrators. They hear everything better than voice and are therefore bothered by background noise. They distinguish between different textures and prefer silk
– Super-vibrators. They hear high-pitch sounds too well and would not mind driving a more silent car like a Tesla. They are very sensitive to touch and prefer to wear cotton and if possible seamless socks

The 3-min Mozart requiem "Tuba Mirum" is a fun way to identify a vibrator profile. The tune starts with a very low tuba bass, around 16 Hz (like a tiger roar), and then suddenly turns into a high-pitched singing voice, creating a pilo-erection—nothing dirty here, just chills making your hair rise. Non-vibrators will not experience any pilo-erection, medium-vibrators will experience one pilo-erection per minute, and super-vibrators will experience one or more pilo-erections per minute (Table 1.2). Experiments

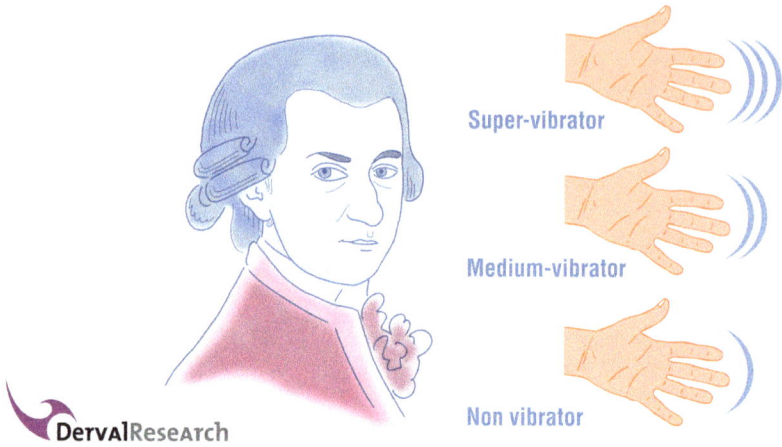

Fig. 1.5 Vibrator profiles (printed with DervalResearch permission)

Table 1.2 Preferences by vibrator profiles (printed with DervalResearch permission)

	Non-vibrator	Medium-vibrator	Super-vibrator
Sound	Needs to put up the volume to hear bass properly	Has difficulties following a conversation with background noise	Very sensitive to high-pitched sounds
Sensitivities	Nail on a chalkboard	All bass sounds	Baby crying, cutlery, klaxon
Favorite instruments	Guitar, saxophone	Piano, Violin	Bass, cello
Touch	Unable to distinguish fine textures under 200 μm. Prefers slick surfaces	Able to distinguish between various textures. Prefers smooth surfaces	Very sensitive to textures. Prefers fluffy materials
Favorite textures	Leather, wool, and silk	Silk and cotton	Cotton
Mozart Requiem "Tuba Mirum" Test	No piloerection	Less than 1 piloerection per minute	1 piloerection per minute and more
Estimated population	25%, more men	50%, more women	25%
Luxury brands	Harley-Davidson, Jaeger-LeCoultre	BMW, Zegna	Tesla, Blacksocks

Based on measurements and observations conducted by DervalResearch on 1200 consumers between 2007 and 2010 (Derval D, Hormonal Quotient and Tactile Sensitivity: a Segmentation Model to Understand and Predict Individuals' Texture Preferences based on Prenatal Exposure to Hormones. *Proceedings of Society for Behavioral Neuroendocrinology 15th Annual Meeting, 2011*, Queretaro, Mexico, p. 125; Derval D., Fingerprint and sound perception: a segmentation model to understand and predict individuals' hearing patterns based on OtoAcoustic Emissions, sensitivity to loudness, and prenatal exposure to hormones, 2010) conducted confirmed also that adventure and sensation seekers were more likely to be non-vibrators (Grewe, Nagel, Kopiez, & Altenmüller, 2007).

Let's see how brands like Harley-Davidson, Jaeger-LeCoultre, and Tesla manage to bring real "sense-appeal" to personas with different vibrator profiles.

1.4.2 A Signature Sound: The Harley Davidson Case in India

Legendary brand Harley Davidson recently decided to revamp its image by entering new markets and targeting different personas like, for instance, women. Very different indeed from the original tattooed and bearded US biker persona. The brand arranged Garage Parties, gatherings where women interested in biking could meet and greet and try out some Harleys, and hopefully be convinced to buy one. The Garage Party campaign was carried out in different countries and within a couple of months, Harley Davidson became the #1 big motorbike brand among women (Derval, 2016). Incredible. The question is how come? It is a good question to ask especially if you are a competitor like Honda or BMW. If the success is due to the Garage Party campaign, then let's just double the advertising budget and beat Harley in that field. So, why do you think Harley became the most popular big bike among women?

The answer is in the product features themselves—as always, from our observations. Harley Davidson became a legendary brand in the US with its distinct sound and cool low seating. The particular design of the motorbike made it very popular among women and "vertically challenged people" as it's the only big bike that enables their feet to touch the ground—very handy for keeping balance on a motorbike. The brand now wants to broaden its customer base by opening it to environment-conscious riders—Harley is right now testing the LiveWire, its first electric motorbike, in Europe and in the USA. The Milwaukee brand also decided to target personas in other geographical areas.

And if we consider the motorcycles' market growth and size, the geographical area presenting a strong market potential is the Asia Pacific. No wonder Harley Davidson decided to enter the Indian market so far dominated by cheaper 100cc and 150cc motorcycles from brands like Hero and Honda. The Indian market is attractive but, in addition to being dominated by cheaper bikes, it presents also strong barriers to entry, notably in the form of taxes. Fortunately, Harley Davidson managed to overcome this major issue by deciding to assemble the motorcycles locally. This strategic move helped decrease the sales price by 40%. An analysis of the luxury market in India revealed an untapped potential of 600,000 millionaire households, eager to show their status by investing in fine dining, jewelry, cars, and why not motorbikes?

It was a smart move as more and more Indians are buying high-end motorcycles like Harley Davidson, Yamaha, or Ducati. There is an intense rivalry though among the players with the rising threat of more local stylish bikes. The dealer's network is widespread. Harley Davidson invested in the relationship with the distribution channels, as it is key to keep control over it when going abroad. Training the staff played also a huge role in conveying the experience. To avoid a "distribution trap" effect, Harley Davidson took care of connecting directly with their Harley Owners Group also called HOGs, building a strong community.

Harley Davidson clearly understood their business is not about motorcycling from A to B, but about how you enjoy the ride. Since its inception in 1905, the brand was associated with motorbike racing and each advertisement was like a mini story, telling how Harley Davidson makes your journey fierce and awesome. Harley Davidson was conveying values similar to Red Bull at that time and linked to recklessness and sensation seeking. Within the heavyweight motorbikes market that includes traditional bikes, sport bikes, and dual bikes for off-road use, the Milwaukee brand is playing in the field of cruiser and touring bikes. Those sub-segments focus more on styling, customization, and traveling, with for instance large luggage compartments or saddlebags.

In the Indian market, the leisure-riding aficionados can be found all across the country. For instance "at 69 years old, Govind Mohanty could be one of the oldest motorcycle riders in India, but that didn't stop him from taking a long trek to the closest Harley-Davidson dealer: in the eastern metropolis of Kolkata, 450 km away from his hometown in Bhubaneswar. Mr. Mohanty did sell his 9500 USD Harley Super Low in June but only to buy himself a bigger bike, the 24,000 USD Fat Boy—the model made famous by Arnold Schwarzenegger in 'The Terminator'" (Derval, 2016).

To increase the affinity, some brands also use brand ambassadors, as we will discuss in more detail in Chap. 6. Bollywood actor Salman Khan is the ambassador of Suzuki bikes and proud owner of a special edition of the GSX-R 1000Z. The Japanese brand even launched the 'Suzuki Biking Lords', a VIP superbike club moderated by Suzuki Motorcycle India Limited (SMIL).

The motorcycles sales represent 76% of the brand revenues, parts and accessories 17%, merchandise including the MotorClothes line 6%, and licensing from Las Vegas chips to lighters 1% of the revenues.

Harley Davidson managed to grasp opportunities in India such as customization. Like Mr. Mohanty, many Indians prefer to buy an entry model and then do some tuning on it, as Bike India journalist confirms when reviewing the launch of the Harley Davidson basic motorcycle, the Street Bob: "In the end I would simply say that Harley-Davidson India has done a fabulous job of customizing this baby. The numerous options that can be purchased to customize your bike in your particular way is a unique concept in India that is only offered by Harley-Davidson for now."

Harley Davidson is a lifestyle brand proposing a unique and augmented experience. Its ability to adapt to local customers and build a strong community was instrumental in India, but not as much as the proportion of non-vibrators, way higher in the country than in let's say neighboring China (Derval, 2010).

1.4.3 Luxury Watches and Other Complications: The Jaeger-Lecoultre Case

Since mobile phones can tell what time it is, we objectively do not need a watch anymore. However watches, and I would say especially luxury watches, are selling very well. Some might argue that the luxury watches business is less flourishing and it is true that anti-corruption laws in some countries like China have had a serious impact on the sought-after wristband corporate gift. Also, more Western young professionals value vintage watches that they consider as warmer and soulful. Not to mention those addicted to connected watches like the Apple Watch or Fitbit, when not combining luxury and technology with a connected Rolex or Cartier (Maillard, 2016).

In the meantime, in India, the young generation is developing an interest in premium and luxury watches, that translated into a good 17% increase in sales, reaching a volume of 1.3 billion USD (Euromonitor International, 2016).

So what is the fundamental need fulfilled by a luxury watch: Being on time? Being on time and stylish? Probably more like it. Also, depending on the context as listed above, it can be a great corporate gift fulfilling the need to impress and be popular, or a great show-off piece helping to fulfill the need for status and getting a promotion or targeted gender attention. It reminds me the funny story of the journalist who borrowed pricey watches and got tremendous attention from otherwise distant air hostesses and hotel receptionists.

Let's find out what makes some watches so expensive and some brands like Rolex, Omega, Jaeger-LeCoultre, or Cartier so successful. Luxury watches are all about complications.

First worn by aircraft pilots and racing drivers, leading brands are competing in the quest for performance, the culminating point being in 1969 for Omega to be the first watch on the moon, with its Moonwalker model. Today it is all about reaching Mars, with brands like Tag Heuer teaming up with Tesla and hoping for the best coming from the Musk group also involved in space research with SpaceX. Target customers are fascinated by technical prowess and complications.

Complications are all about added value and extra features, other than giving the time in terms of hours, minutes, and seconds. You can find basic complications such as day time, time zone, or a stopwatch in most luxury watches, but for grand complications you will need to open your wallet a bit bigger. Grand complications are more complex and therefore available only on pricey models. They include a tachymeter to measure the speed of the wearer, the extremely rare and difficult to build in tourbillon that improves the watch's sense of gravity and precision, or moon phase information, to know when is the auspicious time to get your hair cut (Nuncio, 2013).

Luxury watch lovers have a fascination for complications but that's not all. The secret ingredient is vibration. Similarly to when car buyers compare the performance of the engine, watch lovers evaluate their coveted gems in terms of vph, or vibrations per hour, and the higher the vph, the more expensive the watch. Most movement watches will tick 8 times per second (that corresponds to 36,000 vph), cheaper watches 5 times (or 21,600 vph), and watches with a 5-figure price tag 10 times per second (or 36,000 vph), for a greater accuracy. The sudden interest for vintage movement watches can also be explained by the fact many of them from the 60's or the 70's have 36,000 vibrations per hour for a price under 1000 USD.

Fascinatingly many watch lovers prefer the 8 Hz sound of an automatic movement watch, ticking 8 times per second to a 1 Hz quartz watch, ticking 1 time per second. You can spot the difference as movement watches' seconds turn in a continuous movement while quartz watches' seconds move from second to second. Quartz watches work a bit differently: the pressure on a quartz stone creates electricity (this is called piezoelectricity) and depending on the shape of the stone, a vibration in a certain frequency (usually 32 kHz and it can go up to 2.4 MHz for marine chronometers). Like movement watches, quartz watches come in various shapes, frequencies, and prices as some brands like Seiko or Cartier do a better job at managing the behavior of quartz crystals under various temperature conditions in order to maintain a perfect performance. The purist watch lover would still argue that movement watches are superior to quartz watches: "One is obvious, if you are going to travel to Mars or even further, your space ship eventually loses signals from Earth GPS or atomic broadcasters. High energy cosmic rays can destroy the electric circuits of your watch." (Kpacotka, 2015)

Product preferences in luxury watches will depend on the perception of sound and vibrations, and also the search for performance. Jaeger-LeCoultre built its name by creating the flattest, then the smallest, mechanical movement and has been admired since its inception in 1833 for being a real "manufacturer", with over 400 patents, producing the watch movements internally in the Vallée de Joux in Switzerland and even selling them to other brands like Vacheron, IWC, Patek Philippe, and Audemars Piguet—probably the reason why luxury group Richemont added the brand to its portfolio in 2000. A good move as the brand keeps on innovating with the Reverso, a flip-over watch with a different time zone on each side (Jaeger-LeCoultre.com, 2017).

The Jaeger-LeCoultre masterpiece is probably the 2.5 million USD Hybris Mechanica a Grande Sonnerie. This elegant watch made of 18-carat white gold and 27 complications such as a retrograde perpetual calendar, not one but two mechanisms, and a tourbillon movement that defies gravity and helps ensure highest performance, and can even play 3 variations of the Westminster Carillons melody—how cool. And all this fits in a 42 mm diameter case with a minimalist 11 mm height (Mr Watch Guide, 2013).

Jaeger-LeCoultre is all about performance and authenticity. In its recent campaigns, the brand showcases champions like Polo star Eduardo Novillo Astrada, and also film director Carmen Chaplin, or DNA researcher Dr. Craig Venter, in order to connect with different consumers, and it works (Sorin, 2015). For watch lovers annoyed by any ticking sounds, the solution is a diving watch as they usually have a thicker case that will muffle the sound or, of course, a digital watch. Except for the Master Compressor maybe, Jaeger-Lecoultre has the reputation to be loud and would not be the first choice for medium or super-vibrators (Watch u seek forum, 2017).

While some brands like Breitling are more focused on pilots, Jaeger-LeCoultre sends strong signals in direction of sailors and nature lovers, by partnering with UNESCO in order to help conserve around the world 47 marines sites listed as world heritage (Jaeger-LeCoultre.com, 2016). So if Omega went on Mars, what can Jaeger-LeCoultre associate itself with in order to convey the "manufacture" excellence? Teleportation of course. In the last Marvel movie, Doctor Strange is wearing a Jaeger-Lecoultre Master Ultra Thin Perpetual while traveling from one place to another.

1.4.4 A Silent Success: Tesla Motors

Elon Musk, already known for creating PayPal, dared to design a luxury electric car. Even the popular magazine Motortrend acclaimed Tesla as the car of the year. The California-based electric sports car brand clearly outshined the Nissan, Toyota, and Chevrolet trying to convert drivers to electric cars with squeaking boxes on wheels. Musk was the first to believe in a winning combination of elegance and modernity. With his involvement in space travel, solar energy, and more recently artificial intelligence ethics—maybe to make sure cars do not start biting their drivers?—he became a new opinion leader, joining the ranks of Richard Branson of Virgin and Jack Ma of Alibaba. People are so fascinated by his story that he gained the surname of "real life Ironman" and he was even featured in the movie.

Tesla understood the design that would make a car dreamy like, for instance, the Lamborghini style scissors doors, that go up like some cool insect wings instead of sliding to the side like that of a defeated animal. Putting those doors on the Tesla Model X brought a new crowd to the Tesla showrooms: a crowd including spouses and children and those more serious about buying. Tesla made luxury features—note that this type of door opening is also way more practical on top of it—affordable and environmentally friendly. Like you are showing off but for the greater good.

Only one thing was still missing on the Tesla, or maybe two. First is the vibration, of course. Many are puzzled by the lack of interest for electric cars. You could drive an electric car, if you just need to go from A to B. But what about if your real benefit is to vibrate. You want to vibrate from A to B. Think of your colleagues always eager to drop you home—those without creepy intentions I mean—do you really think they want to win a popularity contest? No, they just want to drive, or more precisely to vibrate. Luckily some third-party application addressed this critical unmet need, and using the accelerometer and GPS of the driver's iPhone, plays the appropriate noise directly from the Tesla sound system, at the right moment and even at the right frequency, as you can choose between the sound of an aggressive car, for non-vibrators, or a motorbike, an airplane, or, how well-thought, a spaceship. When accelerating up to 120 mph, the Tesla S sounds like a laser gun, rather than the usual roaring noise of a luxury car, ideal for super-vibrators (Jalopnik.com). The second miss is the absence of gasoline or petrol smell—as we will see in Chap. 6, some are addicted to it. There's always the possibility to hang a gasoline-scented tree in the car though (Insideevs.com, 2016).

Tesla natively appeals to super-vibrators and can attract other luxury personas with the help of customization.

1.4.5 Brand Naming: The Ermenegildo Zegna Case

Sensory research on brand logo and packaging shows that a congruent combination of sound and shape in the brand name or messaging can enhance the product experience and better address consumers' expectations (Spence, 2012).

Branding is not just about designing a nice logo. Branding is actually called customer relationship. With the help of WeChat interactions, involvement in good causes and various activities, firms can create and nurture their relationship with customers. It is true that many people just show off their Louis Vuitton bag for the logo. And the Bernard Arnault brand must know something about branding as they now have stores from France to Mongolia.

Creating a brand personality with specific codes helps establish this connection. It is not a surprise if many of the legendary brands like Michelin—we will study later in this book—are personified by a character, in this example the Michelin Man, or by a celebrity as we saw with Salman Khan in the Harley case. A good brand name also helps connect with the customers. It is true that many famous brands have complicated names. We can think of Swarovski or Ermenegildo Zegna. And it didn't prevent them from becoming successful. So what is a good name?

A good brand name is a name that sounds appropriate and speaks to the customers. Take Ermenegildo Zegna: it sounds Italian, rounded, sophisticated, and appeals to customers looking for exclusive and smooth-tailored Italian suits, medium to super-vibrators, sensitive to fabrics, or short-armed business men needing tailormade suits. Take Swarovski: it sounds sharp like a crystal and cold like a piece of sparkling ice.

In luxury, the first choice is often a family name. As long as it sounds consistent with the brand codes then it's fine. We have many examples like Chanel, or Dyson. If the family name is not suitable, a first name can do like for example Tiffany, Maxim's, and Mercedes. If it is linked to the story telling, it is even better, like in the case of Mercedes being named after the daughter of the founder. The name of a god or from mythology is also nice, like Hermès, god and messenger. In many cases luxury brands combine the first name and the family name. We have Ermenegildo Zegna, Louis Vuitton, Christian Louboutin.

The name can also refer to a place, combined with a name—think of Shanghai Tang. It can also be linked to the products like Blacksocks.com. Or be a mix between a place and an activity like in the case of BMW "Bayerische Motoren Werke".

If you know your target customers, your brand story, and your brand codes—we will study a helpful framework for that in Chap. 4—the brand name will be an obvious and natural choice.

1.4.6 Vibrator Profiles: Business Applications

Now that you are aware of the critical role of the sense of vibration in male-to-male competition and luxury, the endless business applications include:

- Selling ornaments for males—think of Rolex or the golden iPhone
- Fine-tuning the frequency and intensity of vibrations to produce a vavavoom effect
- Designing the portfolio in luxury strata (premium, luxury, super luxury) to allow ranking among peers

- Including a "black & gold" or "platinum" collection for the pack leaders
- Choosing a brand name that sounds good to the target customers
- Targeting a country with vibrator profiles matching the product attributes

1.5   Profiling Luxury Shoppers

In this section we will see how to profile luxury shoppers using the persona framework, learning from the best like BMW and American Express.

1.5.1   Profitable Customers: The American Express Case

Credit cards are very effective ornaments and, believe it or not, their chipset is made of gold. You might somehow regret you threw your expired cards away.

Let's take a closer look at the credit card business, characterized by selected customers, transaction fees, and differentiated business models depending on the target customers.

American Express is a credit card acting as a financial service and competing therefore with WeChat and similar mobile applications that enable instant payments. It can also be considered more broadly as a premium service. Competitors might then be loyalty cards from luxury brands, also offering special deals and events and payments facilities.

The tricky part in the credit card business is that "you want to sell but not to anybody". The key is to monitor the risks. In the USA for instance, most credit providers will check the following criteria in order to grant access to a credit card: payment history, amounts owed, length of credit, new credit accounts, types of credit in use (Derval, 2016).

In the USA, a top card issuer like American Express, leading with almost 25% market share, gets most of its revenues from transaction fees.

The business model in the premium segment occupied by American Express consists in providing credit cards and "a great service to fewer customers making larger transactions". In the standard segment, the debt or credit cards are often linked to the bank and cover many small transactions. American Express decided, from the go, to focus on high-net-worth individuals, and developed a cutting edge customer service linked to their social media and various touch points.

The recession created a new behavior among card issuers and they tended to favor risk reduction. Most USA credit firms for instance are now focusing exclusively on good risks—consumers with a credit score above 720, knowing that the average nationwide is around 660.

The American Express strategy is to remain relevant and accessible to a new generation of high-net-worth customers by setting up social media as the primary customer touch point and adapting their CRM model to local sensitivities and standards. They understood that branding is all about consumer experience and how to deliver successful integrated communications via various touch points.

And they prefer to recruit new affluent customers in new regions such as Asia than to target middle-income customers.

Up-selling and cross-selling new products and services to existing customers first is way more cost effective. Also, the same type of customers can be handled by the same trained team whereas a new customer segment often implies new organization, infrastructure, and people.

The key recommendation would be to focus on the right persona and develop cross-selling and up-selling rather than having too many irons in the fire.

Identifying the target persona and right positioning—we will study the second one in Chap. 3—are the two essential steps of any luxury marketing strategies. Once these are clear, everything from writing a marketing plan, a business plan, or an opportunity study becomes a simple formality.

1.5.2   A Seamless Service: The Blacksocks Case

Let's have a closer look at what luxury shoppers really want, with the Blacksocks case.

Samy Liechti, the general manager of Blacksocks.com, built an empire upon a rather embarrassing anecdote. While he was still a young manager in Switzerland, he had to attend a very important business dinner. He put on his nicest suit—with stripes, of course—as he was a consultant. When he arrived, he discovered that the restaurant was Japanese and they had to take off their shoes. This was a big moment of embarrassment when he realized that one of his socks had a hole and that actually the other one was more dark grey than black. Doing a quick market research, he found out that most men do not want to waste time buying socks. Also, the assortment in shops rotates often. So even if you find the ideal pair of socks, it will soon be replaced. And nobody wants to spend the weekend sock hunting.

So based on this opportunity, he designed the Sockscription, in which every quarter, clients receive by mail three to four pairs of socks—a little bit like they would receive a bottle of milk or in ancient times the newspaper. So far, no competitor has been able to measure up with Samy because he is not only sending socks but he created a luxury service.

Instead of attaching an invoice to the socks, he is sending a letter with useful advice on etiquette, and how to match your socks with your suit and what types of ties to wear. All very useful information for his target business customers. The customer service is always available and any question is considered of the highest importance. Thirty percent of the new customers are recommended by existing customers. And 25% of the customers became brand ambassadors. After a successful launch in Switzerland and in Germany, Blacksocks is now targeting the American market.

When Samy Liechti started Blacksocks in 1997, he struggled to convince socks brands to sell online. They didn't believe in the concept, so he decided to create his own socks brand. A couple of years later, the concept is a huge success thanks of course to the quality of the products, but, most importantly to the attached luxury service.

Fig. 1.6 Blacksocks invisible socks (printed with Blacksocks permission)

One of the customers was very happy by the way he was treated by the brand and he happened to be a member of the consumer association in Switzerland. He decided to nominate Blacksocks for the consumer awards and they actually won. This generated a 70% increase in sales. So, great story telling and attention to details pay.

This is also why Blacksocks has been heavily copied—since its inception, 180 companies decided to also sell socks online and to create their own "Sockscription"—but none of them succeeded. Because it is not just about putting socks in an envelope—it is about the very careful aftersales service and the useful advice on styling. Also, Blacksocks is the only brand who managed to make socks exciting (Fig. 1.6).

Using the latest technology, Blacksocks was able to release some exciting and crazy offerings like "smart socks". Thanks to an RFID chip in the socks, you can check on an iPhone app at the exit of the washing machine whether your socks are properly paired or not, and if some are missing. Another innovation is the "blackometer". You can point your iPhone at your socks to identify how black they are and never end up again in the Japanese restaurant situation with a dark grey and a light black sock.

Of course, some people were amused by the application, but again the fact that socks are boring is the reason why businessmen did not want to get involved in the first place. So as soon as it gets a bit more exciting and technological, they get more interested in the product, like in the case of the baby carriage designed by wannabee Formula 1 racing brand MacLaren or the Dyson vacuum cleaner as we will check in the next chapter.

Finding the right activities is important when creating a luxury brand. The big question is what type of activity should the company be doing. Let us take the example of Blacksocks (Fig. 1.7). Ten years ago, Samy Liechti created the company with a

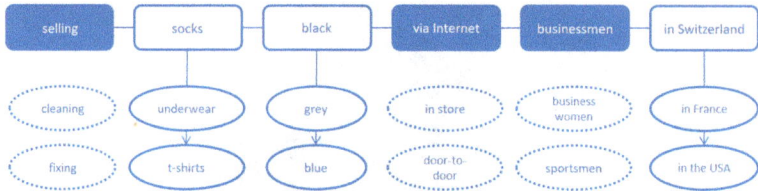

Fig. 1.7   Blacksocks mission (printed with DervalResearch permission)

mission to sell Blacksocks via Internet to businessmen in Switzerland. He quickly realized that his job was to sell socks not to repair or clean or manufacture them.

Why? Because his strong point was the sense of service. He perfectly understood the expectations of the businessmen target. For instance, targeting businesswomen is not in his plans. His business model is based on word-of-mouth. And door-to-door and in-store sales were also not considered.

In order to develop the brand, the team can expand to other countries like France and the US and also to products other than socks: why not underwear or t-shirts? Socks can be available in colors other than black, like grey, or blue, or funky colors. The range successfully extended to sport socks for squash and skiing, as sports and luxury are a good match—we will see why when studying the sense of motion.

For brands, understanding the reasons of their success is the only way to reiterate it. Blacksocks cotton socks are very appealing to super-vibrators, and it happens that you can find more of them in Switzerland or in the US than in France for instance.

Knowing customers at a deep level, including their sensory perception, is key.

1.5.3   The Personas Framework: The BMW Case

BMW is an emblematic car brand, leading luxury car sales around the world. Let's dig into the brand's personas.

BMW stands for "Bayerische Motoren Werke" which means motors from the Bavarian province in Germany. Actually the brand was originally called BFW for

"Bayerische Flugzeug Werke", which means Bavarian aircraft motors, as back in 1916 the firm was working on airplanes.

Always on top of the game, BMW fuels its cars with innovation and proposes a winning combination of aesthetical and practical accessories, not leaving aside its focus: performance. A great example is the night vision system provided by FLIR on most BMW series 7. Using latest infrared technology, BMW Night Vision with Dynamic Light Spot detects people and animals in the dark and sheds light on them. Another very useful feature is to be able to open the trunk of the car just by kicking it with the foot: How awesome! The secret behind it is a laser sensitive to the movement and operating the trunk opening.

Carrie welcomed us at the BMW Brand Experience Centre in Shanghai: "When people pass by the Chinese pavilion, they see this transparent crystal box which is very eye-catching! Our design won three design awards including the Red dot design award", she introduces proudly, wearing her cool BMW Product Genius light blue jacket. In the gigantic showroom you can admire fashionable, futuristic, artistic, and even vintage BMW models, that in some ways shaped current BMW sports characteristics and brand codes. The showroom is intended to raise customers curiosity and make them want to know more about the brand.

First she volunteered to share some secrets about BMW cars design and features particularly appealing to the customers. "Their age might vary but our customers are all young at heart", she explains, "customers choose a specific model depending on the performance. For example, the series 7 is favored by business customers wanting to show personality and status. Some customers use the car for trekking purposes and would opt for our SUV model X4 in a vivid color. Other customers are looking for a practical car and will choose a compact city model like the i3. We offer a wide range of colors selection such as Melbourne red and Snowy mountain white to adapt to customers' mood", explained Carrie as she pointed at a color selector where I counted 12 shades of grey, one black, one white, and one red option—we really don't have the same definition of colors—we will see in Chaps. 3 and 4 where it comes from and what the implications are when designing luxury brands.

For now, we are ready to do a test drive and Carrie is our pilot—as I write our cameraman is still recovering from the driving experience. "Let's do a linear acceleration. This car accelerates at the speed of 7.9 s", emphasizes Carrie enthusiastically. "Although i3 is an electronic car, you can still have the experience of driving a gasoline car. You can also feel its strong sense of motion. It reacts very fast whenever I steer the wheel" she continues, talking while piloting like an F1 racer. Carrie explained that the interior decoration is made of environmentally friendly materials, like olive-oil tanned leather, so that you do not have the typical new car smell inside—a heaven for super-inhalers as we will see in Chap. 6.

Usability and driving experience are key at BMW. They designed reverse hinged rear doors for an increased passengers' convenience, for instance (Fig. 1.8). They know their customers and we will analyze a typical customer, Gary, with the persona framework in following pages.

Initiated by designers fed up with vague and dreamy aspirational customer segments, the persona framework is a visual and powerful tool to effectively profile

Fig. 1.8 BMW i8 (printed with BMW permission)

customers, and is part of the Design Thinking approach. Now it is more broadly used by leading brands and across company departments. Let's study Gary and see how the persona framework can help put a face on a luxury car.

The luxury goods sold worldwide are mainly purchased by Chinese consumers, particularly excited by powerful cars and luxury fashion items, followed by Europeans, Americans, and Japanese shoppers. A report identified luxury consumer segments in the Chinese market such as the omnivore, with an average yearly luxury shopping budget of 23,000 rmb (3000 USD), mostly buying jewelry and watches from well-known brands while traveling, and based in tier two and three cities (Derval, 2016).

Like most marketing segments, this is a bit vague and difficult to manipulate. How do we recognize these people in the street? What does an "omnivore" look like, no but seriously? Where can we find them? Who do they hang out with? How shall we approach them? Why do they behave like that? We will see now how to apply the persona framework in order to identify and describe target luxury customers in a more effective way.

A persona is a representative customer, inspired by existing customers, with a clear motivation, job, hobbies, and lifestyle, someone that you can identify in your network, that you can recognize in the street. Think of the tuhao described in our golden iPhone case (Fig. 1.1). The persona is so true that he will inspire us for the creation of new products, of new potential markets, and for finding innovative distribution channels.

Fig. 1.9 BMW Persona Gary (printed with DervalResearch permission)

The persona technique is used by successful companies, like Apple or Philips, and will help us identify the profile of target customers, their expectations, their interests, and their preferences. It is a powerful tool to describe target customers' segments, whether they are consumers or deciders within the company.

To describe a persona, we would start by a little drawing and document the following questions.

What are their motivations in life?

What is their age and typical salary? What do we know about their job? Hobbies? The type of studies they did ? Do they have a family? Children? Pets? Do not hesitate to complete with questions more specific to your product or project.

To properly describe a persona, the idea is to focus on common points and occurrences. For instance if two business consultants are called Gary, let's call him Gary. If most of them are around 34 years old, let's put 34, even if some are 25 years old and some 48. The point is to avoid averages and intervals and pick instead the most frequent features in order to get a clear and accurate picture. You know your persona is successful when people in your extended network look at it and say "I know this guy!".

Adding information about the perception is very helpful. Especially aspects related to the luxury product or service you consider. Usually a given brand has 7–8 Personas, and they each have specific needs and preferences.

Let's meet Gary—an e-business consultant and happy owner of a BMW (Fig. 1.9). Gary is all into performance, delivering the best, the fastest. That is why he enjoys the fast-paced e-business environment and is into technology. When not piloting his car—he picked a consulting job to be able to drive for work every day—he plays racing games on his Xbox, when his wife permits! In between appointments, he loves to check the latest technological gadgets online while enjoying a strong coffee at a café terrace.

His Hormonal Quotient® is balanced—we will see that in more detail in Chap. 5. He is very outgoing, and has a slight tendency not to respect the speed limits. Gary is into status, technology, and speed, and would be driving a BMW Series 3 or a BMW i8 if the business is going well. Drivers who buy a BMW i8, or a Series 3 can be very different from the ones buying a BMW i3, or X4.

Knowing the persona helps with developing the right features. For instance the iDrive, a unique button commanding all the functions inside BMW cars, was a very controversial feature at launch because many journalists argued that it was way too complex to use. When observing that a BMW persona like Gary used to play with his complex Xbox console, it is obvious that the iDrive is like a walk in the park for him.

1.6   Take-Aways

Male-to-Male Competition

- Male-to-male competition feeds the luxury industry
- Ornaments that will help impress opponents are in luxury shoppers' top purchases, from BMW to golden iPhones

Vibration

- Sound and touch are all about vibrations and key to luxury shoppers, from cars to watches
- A luxury brand name should sound good and be congruent with the brand codes

Personas

- Usual marketing segments by age, gender, or revenue are too vague
- The Persona framework helps identify target luxury customers in a more effective way

- Knowing the Persona will help develop the right luxury products, services, features, save on the advertising budget, and connect the brand with the right luxury shoppers

References

Acharya, S. (2016, March 29). Gold iPhone SE drives 3.4 million pre-orders in China. *International Business Times*. Retrieved October 15, 2016, from http://www.ibtimes.co.uk/gold-iphone-se-drives-3-4-million-pre-orders-china-1552043

American Institute of Physics – Inside Science News Service. (2000, December 29). The secret of a tiger's roar. *ScienceDaily*. Retrieved January 30, 2018, from www.sciencedaily.com/releases/2000/12/001201152406.htm

Bergman, J. (2010, June 30). China's TV dating shows: For love or money. *Time*. Retrieved October 15, 2016, from http://content.time.com/time/world/article/0,8599,2000558,00.html

Bergstrom, M. (2012). Tipping gender scales: From boys rule to girl power. In *All eyes east* (pp. 65–89). New York: Palgrave Macmillan US.

Cameron, L. A., Meng, X., & Zhang, D. (2016). *China's sex ratio and crime: Behavioral change or financial necessity?* (IZA DP No. 9747). Available at SSRN 2737718.

Derval, D. (2010). *The right sensory mix: Targeting consumer product development scientifically*. New York: Springer.

Derval, D. (2016). *Luxury brand marketing 奢侈品品牌营销:创建·实施·案例*. Shanghai: Donghua University Publishing.

Derval, D., & Bremer, J. (2012). *Hormones, talent, and career: Unlock your Hormonal Quotient®*. New York: Springer.

Euromonitor International. (2016). *Country report: Watches in India*. Retrieved from http://www.euromonitor.com/watches-in-india/report

Farley, S. (2014, July). *A brief tour of automotive sound sources*. Designing Sound. Retrieved from http://designingsound.org/2014/07/a-brief-tour-of-automotive-sound-sources/

Grewe, O., Nagel, F., Kopiez, R., & Altenmüller, E. (2007). Listening to music as a re-creative process: Physiological, psychological, and psychoacoustical correlates of chills and strong emotions. *Music Perception, 24*(3), 297–314.

Hurun. (2014). *Wealth report 2014*. Hurun.

Jaeger-LeCoultre.com. (2016, July 6). *Jaeger-LeCoultre highlights its partnership with the UNESCO World Heritage Marine Programme: Preserving the heritage of time*. Retrieved from http://www.jaeger-lecoultre.com/au/en/chronicles/news-events/unesco-world-heritage-marine-programme.html

Jaeger-LeCoultre.com. (2017). Retrieved from http://www.jaeger-lecoultre.com/us/en/home-page.html

Jirotkul, M. (1999). Operational sex ratio influences female preference and male–male competition in guppies. *Animal Behaviour, 58*(2), 287–294.

Kpacotka. (2015). *Reddit*. Retrieved from https://m.reddit.com/r/Watches/comments/31hshv/question_high_end_quartz_watches/

Loveday, E. (2016). With app, Tesla Model S sounds like a muscle car. *Inside EVs*. Retrieved from http://insideevs.com/app-tesla-model-s-sounds-like-muscle-car/

Maillard, P. (2016, July 20). The state of the luxury watch industry in 2016. *Luxury Society*. Retrieved from http://luxurysociety.com/en/articles/2016/07/the-state-of-the-luxury-watch-industry-in-2016/

Math, D. (2013, July) The sound of Formula-1 cars played on the guitar. *Total Car Magazine*. Retrieved from http://m.totalcarmagazine.com/features/2013/07/07/the_sound_of_formula-1_cars_played_on_the_guitar/

Mr Watch Guide. (2013, March 6). *The most complicated wristwatch yet by Jaeger LeCoultre*. Retrieved from https://mrwatchguide.wordpress.com/2013/03/06/complicated-watch-by-jaeger-lecoultre/

Niedermeyer, E., & Lopez da Silva, F. H. (2004). *Electroencephalography: Basic principles, clinical applications, and related fields*. New York: Lippincott Williams & Wilkins.

Nuncio, S. (2013, March 8). A watch complication doesn't have to be complicated. *Jonathan's Watch Buyers*. Retrieved from http://www.jonathanswatchbuyer.com/2013/03/watch-complication/

Prigg, M. (2015, May). How to speak HORSE. *Daily Mail*. Retrieved from http://www.dailymail.co.uk/sciencetech/article-3083762/How-speak-HORSE-Researchers-equines-express-emotion-whinnies-reveal-frequencies-used.html

Sorin, K. (2015, May 15). Jaeger-LeCoultre campaign highlights diverse lifestyles to connect with consumers. *Luxury Daily*. Retrieved from https://www.luxurydaily.com/jaeger-lecoultre-campaign-highlights-diverse-lifestyles-to-connect-with-consumers/

Spence, C. (2012). Managing sensory expectations concerning products and brands: Capitalizing on the potential of sound and shape symbolism. *Journal of Consumer Psychology, 22*(1), 37–54.

Watch u seek forum. (2017). *Wristwatches and ticking*. Retrieved from http://forums.watchuseek.com/f2/wristwatches-ticking-search-silent-wrist-watch-952444.html#/topics/952444

Zhang, R., Pian, H., Santosh, M., & Zhang, S. (2015, March). The history and economics of gold mining in China. *Science Direct*. Retrieved October 15, 2016, from http://www.sciencedirect.com/science/article/pii/S0169136814000572

Zuk, M., Thornhill, R., Ligon, J. D., Johnson, K., Austad, S., Ligon, S. H., et al. (1990). The role of male ornaments and courtship behavior in female mate choice of red jungle fowl. *American Naturalist, 136*, 459–473.

Identifying Profitable Markets

2

> *"You shouldn't be that ambitious. Set a small target first,*
> *like earning 100 million yuan (15 million USD)."*
> Wang Jianlin, China's richest man (Dan, 2016)

Luxury markets are vast and it is easy to get lost in the wrong country or with the wrong target persona, as we will see with the superyacht case. Also, success can be tied to a specific geographical area or luxury shopper, like for the city of Dubai and Moncler. And the benefits framework will help us identify these profitable markets.

2.1 Introduction

In this chapter, we identify profitable markets for luxury brands and clarify what exactly luxury shoppers and especially Ultra High Net Worth Individuals (UHNWI) want in Sect. 2.2 with the superyacht case and answer following questions:

- How to identify profitable markets for specific luxury products and activities, like yachting and skiing?
- Why are some products more successful in certain geographical areas than in others?
- What can make or break a purchasing decision?
- What are luxury shoppers' drivers?
- How to appeal to UHNWIs?
- How to turn product features into customers' benefits?

We see in Sect. 2.3 that status-seeking is in fact physiological and therefore more localized in certain areas of the world.

© Springer International Publishing AG, part of Springer Nature 2018
D. Derval, *Designing Luxury Brands*, Management for Professionals,
https://doi.org/10.1007/978-3-319-71557-5_2

In Sect. 2.4, we reveal that the sense of motion is driving the whole luxury industry, what the sense of motion exactly is, and we check the success stories of Moncler and Moutai.

In Sect. 2.5, we learn how to systematically turn product features into customers' benefits through the Stella McCartney and the Dyson cases.

Essential learnings are grouped in the Sect. 2.6.

2.2   What Do Luxury Shoppers Want? The Superyacht Case

A friend reminded me of the successful French advertising campaign "Au revoir Président" ("bye-bye CEO") showing in a selfie-like video a guy dressed in holiday-cruise mode entering the board room being abnormally chilled and relaxed, and interrupting the meeting to say bye-bye to the CEO, who is visibly annoyed. The relaxed guy just won the Lottery and did what many people dream of: he quit. One would think that the natural next step would be to buy an island in the sun and a yacht. As we will see in this yacht case, this is not true for everyone.

2.2.1   Welcome On-board

The team arrived on-board—4 engineers. They were asked to follow the steward to the master living-room, the sheikh would receive them in a minute. The Norwegian engineers were stressed and excited at the same time. It was the first time they would meet one of their big clients. Also, even though they were designing accessories and equipment for yachts, it was the first time they actually had a chance to visit a superyacht. It was also the first time they had visited Dubai, and they were surprised by the alignment of luxury cars coming in crazy colors as well as the abundance of branded accessories and luxury resorts and restaurants. No wonder Emirates airline became so popular with their wider seats and full-monty entertainment system, they thought. Not to mention the high towers they were planning to visit later that day if possible.

The almost 200-m long yacht was managed by a busy crew of 10 hostesses and stewards and piloted by a captain. At least 80 people could be welcomed aboard without even feeling too cluttered. One of the engineers came back from the toilets and was just recovering from seeing so much gold at once while the others were guessing how many hundred inches the big screen in the waiting lounge could be. They took the opportunity to make a last team debriefing, on who would say what and ask what. There was a lot at stake, as the sheikh wanted to buy new piloting equipment for the superyacht and this meant a lot of money.

2.2.2 UHNWIs and Yachting

When Wang Jianlin, China's richest man, was featured on TV and people could hear his nuggets of wisdom, they were quite confused as they didn't expect advice like "set a small target first, like earning 100 million yuans (the equivalent of 15 million USD)". Wang is like Jack Ma from Alibaba and a couple of others, an UHNWI: Ultra High Net Worth Individual. If you wonder whether you are one as well or not, it is simple, check your net assets, and if they are higher than 30 million USD bingo!—you are an UHNWI. Half of them are based in North America, led by Bill Gates, Jeff Bezos, and Warren Buffet, one fourth of them in Europe, and a sixth in China. The U.S., China, and then the UK count the most UHNWIs. UHNWIs are heavily involved in purchases like luxury cars (cars worth over 100,000 USD), superyachts (yachts longer than 24 m), private jets, and collectables (fine wines, jewelry, watches, and arts) (Knight Frank Research, 2016). But the taste in luxury expenses differs depending on the region: UHNWIs in the Pacific region are three times more likely to own a yacht than an American. Middle-Easterners, Europeans, and Latin Americans are also fond of their floating luxury, while Asian UHNWIs would rather spend their disposable millions on other items.

2.2.3 Meet the Sheikh

Finally, the engineers were accompanied and introduced to the sheikh. He was sitting comfortably, enjoying tea with some sweets and in perfect English proposed that the visitors enjoy some refreshments. The engineer in charge of taking notes of the sheikh's requirements just wrote the date in his notebook and was ready to drink his words. The sheikh started "So, what do you propose?" The engineers looked at each other like lost minions, confused. One dared "What do you mean? We are here to write down your requirements so that we can design your piloting equipment." The sheikh took a break. Then he started again "Listen, I have a very big budget, I do not know exactly what you can do. So, you tell me, what can you propose?" Understanding the total shock he put the engineers in, he added magnanimously "Maybe I will let you think about it, and you come back and tell me". Too happy to being given an alternative to staying there like dummies, the engineers replied enthusiastically "Sure, of course, let's do that. You will not be disappointed."

They left the superyacht confused and more stressed than when they arrived. But what did the sheikh mean? What does he want? What shall they propose? They have some standard components they can play with and assemble or customize in different ways but without clear requirements they felt totally lost. "What does the sheikh want?" is a critical question and very often luxury shoppers' needs are not clearly expressed. Investigating human neuro-physiology and status-seeking mechanisms should help us find the answers.

2.3   Neurosciences, Status-Seeking, and Luxury Markets

We have no other choice than buying luxury items. At least some of us have no choice.

2.3.1   The Need for Luxury Is Physiological

Even while doing grocery shopping, people think of their status: What will others think if I buy this brand? Look at me buying this fashionable product or expensive brand. Interestingly enough, 96% of the observed shoppers were convinced that they were being judged by their peer shoppers and the least wealthy ones attached the most importance to how their shopping cart would rank (Palma, Ness, & Anderson, 2017).

Recent research confirmed that some people have a higher binding potential on certain dopamine receptors and are therefore looking more for social status—because showing status is extremely rewarding to them (Fig. 2.1).

For some people, status-seeking is a physiological need, in the same way as eating and drinking.

Sheikh Mohammed managed to turn a land of desert into a luxury and innovation hub within 10 years, following the footsteps of his father Sheikh Rashid, who initiated the change. He highlights that UAE was built on trade, not on oil. A huge fan of equestrian sports, a poet, and an entrepreneur, His Highness (HH) owns 99% of top luxury groups—Emirates Airlines and Jumeirah, to name a few—and championed projects like Burj Al Arab, Dubai Opera House, and the brand-new Bollywood attraction park. Father of 23 children, he is working closely with the crown prince Hamdan, his second son and also a recognized equestrian and poet nicknamed Fazza. Crown Prince Hamdan also loves anything related to racing, from cars to yachts, including drones—he just announced the first international drone race.

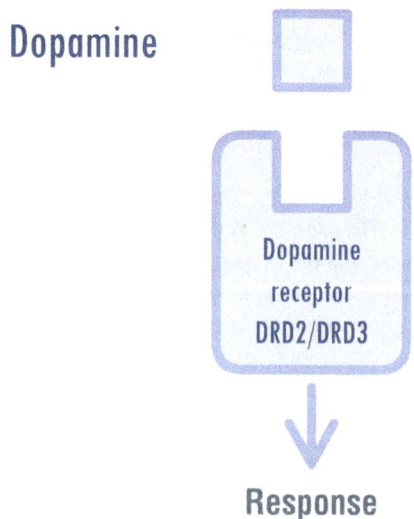

Fig. 2.1 Dopamine DRD2/DRD3 (printed with DervalResearch permission)

The ambitious objective to turn UAE into the place to be by 2021 has been achieved and will be celebrated in 2020 with the international exhibition. Now Sheikh Mohammed, also involved in charity and anti-corruption, wants to put efforts into reaching planet Mars and developing the Emirati space presence.

Sheikh Mohammed also owns the third biggest superyacht in the world, called "Dubai" and able to welcome up to 115 people on its 196 m length (Forbes 2017). Staying true to his Bedouin origins, HH likes to have lunch in the desert every second week. Maybe just contemplating his new target—the sky. Fifty years ago his father Sheikh Rashid released a stamp illustrating a spaceship and today his son dedicates a team of Emirati scientists and 20 billion dirhams (5 billion USD) to build a probe to Mars.

2.3.2 Status-Seeking and Geographical Area

Interestingly, all rich people are not looking for status, and luxury. This becomes particularly obvious when checking countries like Norway. This land of Vikings has the highest income per inhabitant—due to the high concentration of oil—still Norwegians dress up in a very casual way. While enjoying outdoor activities like trekking, men and women wear similar and not very fashionable outfits: jeans and "timberman shirts" or North Face jackets. Most importantly, they do not display signs of status (Derval & Bremer, 2012).

On the other hand, in other countries, people sometimes put their whole salary in a golden or rose gold iPhone. For them status is a need, and luxury products are a must-have. In countries like Brazil, Jordan, Turkey, Italy, and France, the golden iPhone is even used as a currency and when people queue to buy the precious item at the Apple store, the vendor wonders "You want only one?"

When admiring the Chinese Pavilion and the new urban landscape recently erected in Shanghai, I was reminded of France raising its Eiffel Tower for the Paris World Expo. Both countries are similar in many ways: a powerful history and national pride, a centralized power, an elite, an agricultural background, high-speed trains, and a sense for fine foods and fashion. Another critical common point is strong regional disparities.

The French like to show off and are seeking status. The society and companies are very hierarchical and one's power is often assessed by the number of windows in the office, the size of the car, or the number of team members. Even in a romantic context, status is driving the sales. Wai Wong, Category Manager at the prestigious flower group Monceau Fleurs in Paris, confirms that for Valentine's Day the top selling flowers are the biggest ones. In the context of flowers, biggest means the ones with the longest stalk! So no wonder many luxury brands are coming from France—Chanel, Dior, Hermès, Louboutin, Sofitel, Maxim's, Chaumet —to name a few, and more and more are rising from China.

Based on the observation that many tropical animals were more colorful, researchers compared the behavior of the same fish species in different latitudes. Males tend to be more aggressive, into male-to-male competition, active in courtship and females pickier in lower latitudes—understand closer to the equator—where the sex ratio is stronger (Fujimoto, Miyake, & Yamahira, 2015). Could it be that similarly, people in lower latitudes are more into status than their higher latitudes counterparts? If we take our Norway versus UAE example, it is true that Norway has a latitude around 60, and UAE 30, while China or France are in between with 45. A hypothesis to investigate further. What is certain is that bigger cities—a bit like our low latitude waters—attract status-seeking individuals, with different drivers in life.

2.3.3 Luxury Shoppers' Drivers

It is not true that you first need to eat then to buy a designer bag. King Bollywood actor Shahrukh Khan explained during a presentation at Yale University how seeing his parents miserably fail became his main driver in hunting for success and celebrity. He obviously looked at the world through a strong innate status-seeking prism, exacerbated by the negative experience. Other people for instance are more into saving and seeing their family spend outrageously makes them save even more. Still, the saving gene was already present in them.

Unlike many psychologists reducing people to two basic needs like eating and mating, Steven Reiss, professor of psychology and psychiatry at Ohio State University, identified 16 basic desires: power, independence, curiosity, acceptance, order, saving, honor, idealism, social contact, family, status, vengeance, romance, eating, physical exercise, and tranquility. These desires can be found in animals as well, and seem therefore rooted in genetics. Individuals are unique and are not motivated by the same desires. For instance some people are curious and love to learn while others couldn't care less (Reiss, 2004).

The *Fundamental Motives*, proposed by psychologists to explain human behavior, has evolutionary implications, and segments individuals into groups depending on their deep motivation including affiliation (caring for family, making friends), personal safety (evading physical harm, avoiding disease), romance (acquiring a mate, keeping a mate), or status-seeking (attaining status) (Griskevicius, Ackerman, Van den Bergh, & Li, 2011; Griskevicius & Kenrick, 2013).

Luxury consumers are more diverse than the literature might suggest: some individuals will buy an expensive watch to show status, while others will buy a luxury product because nobody else owns it. Motivation is the main segmentation criteria used for explaining luxury consumption, with "bandwagons" of shoppers trying to fit in and purchasing the products on trend or recommended by celebrities and influencers, and "snobs" trying to stand out, and leaning towards rare and contrasting products (Kastanakis & Balabanis, 2014).

According to the costly signaling theory, people engage in activities involving considerable effort, time, or money—like luxury purchases or charity donations—in order to send possible mating signals, displaying desirable attributes like resources or altruism (Griskevicius et al., 2007). Luxury products could be compared to stimuli, provoking positive or negative emotions, depending on the motive of the consumer. A luxury lover could for instance feel rewarded by luxury products, and find them aesthetically pleasing, while a non-luxury lover would find the same products useless (Pozharliev, Verbeke, Van Strien, & Bagozzi, 2015).

The predictive power of the fundamental motives is promising in the marketing field as specific motives lead to specific behavior: status-seeking concerns will lead to higher risk-taking, dating concerns in women will lead to the willingness to be more agreeable, whereas affiliation concerns will lead to conformity and to preferring the most popular products and brands (Griskevicius et al., 2011). Linking neurophysiological variables to the fundamental motives would help elucidate why and how luxury consumers prefer given luxury products.

Based on DervalResearch observations and measurements, status is indeed a huge driver for some. Then what might differ, among status-driven people, is who they benchmark themselves with—all males or females for instance or just peers—and the weapon they decide to use to achieve status: performance (whether scientific, sports, or artistic), popularity (through charity, networking, emotional connection), power (via politics, coercion, sex).

Luxury shoppers can be split into three groups (Table 2.1):

- People driven by performance want to be a recognized champion and they just want to be at their own best. Sometimes they do not even wear so many luxury brands. When they do wear or use luxury items, they are a true inspiration to many followers. People into performance will buy innovative luxury items like well-engineered cars, watches, or jackets like Moncler, in order to support their goals
- People driven by power are the cash cow of the luxury industry as they want to display their power in real-life or on social media. Some are less wealthy yet and will find activities such as blogging or working in the luxury or fashion field that puts them into a power position or in contact with luxury without having to pay for it. Power people will consider the growing profitability as the ultimate proof of success for a luxury brand. Therefore strategies must be well-thought out in order to not negatively affect the overall margins. Power people will buy the most expensive and leading luxury items like Hermès
- People driven by popularity tend to follow people into power or performance and would buy the more affordable model, or they would follow a designer who embodies the causes they and their clan believe in. They would also be more likely to blend in. Popularity people will follow the movement and buy brands like Burberry

Sheikh Mohammed ben Rashid is for instance performance driven and is trying to surpass himself: "True excellence does not come from being superior to others, but means surpassing oneself time and again." (Al-Maktoum & Bishtawi, 2006). Taylor Swift is clearly power driven as we will see in more details in next chapter while Mark Zuckerberg wants to be popular and follows the flow—his Facebook posts focus on good causes and daily life and are always inclusive.

In a country like the United States of America, all types of luxury shoppers can be found in any state, but some concentrations can be observed, corroborated by Cambridge University research investigating Americans' temperament and attitude in each state (Wilson & Kluger, 2013). It's no surprise that most power shoppers are concentrated on the East coast around New York, Boston, and in Texas (see Fig. 2.2), while performance shoppers are more on the West coast, especially around

Table 2.1 Luxury shoppers' drivers (printed with DervalResearch permission)

	Performance	Power	Popularity
Objective	Wants to be at his/her best in a given field	Wants to be feared and respected	Wants to be included and appreciated
Tools	Any luxury and non-luxury products and services that will enhance one's skills and bring extra resources and abilities to complete the posed challenge	Any rare and expensive luxury item or social media channel that will showcase the social rank and dissuade others from messing with them	Any popular luxury item or trend that will confirm they belong to the group and make them likeable
Luxury items	An accurate and versatile mechanical watch for a pilot, a matching jacket for a fashion aesthete	A bag made of a rare animal, the latest designer shoes, a big diamond, a face-lift	A charity ticket, a well-known scarf, an electric car
Luxury brands	BMW, Moncler	Louboutin, Rolex, Hermès	Burberry, Stella McCartney
Celebrities	Sheikh Mohammed ben Rashid	Taylor Swift	Mark Zuckerberg

Based on measurements and observations performed by DervalResearch on 1000 luxury shoppers in over 25 countries from April 2012 to March 2017

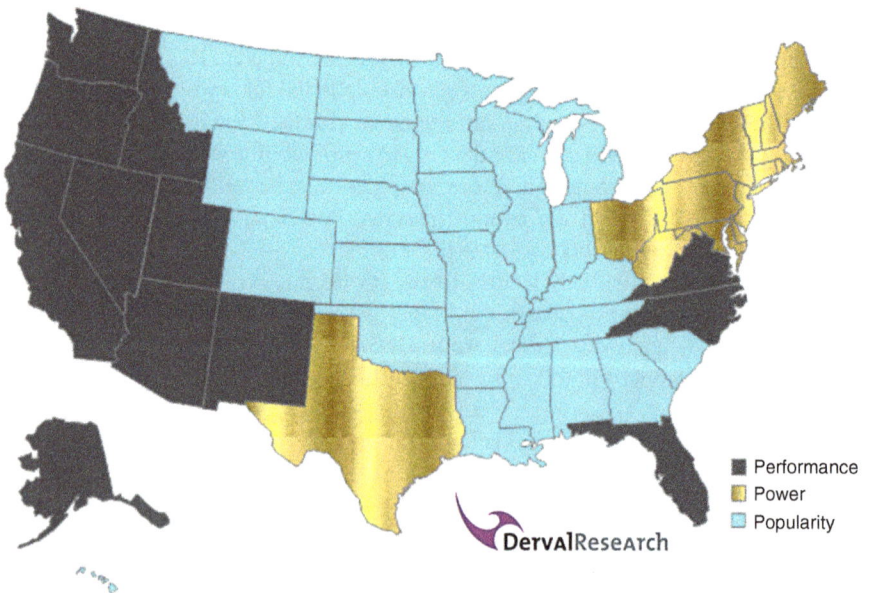

Fig. 2.2 USA map of luxury shoppers motivations (printed with DervalResearch permission, sources: DervalResearch field research, Wilson & Kluger, 2013)

Seattle and San Francisco, and some can also be found near Baltimore and Miami. Popularity shoppers predominate in the vast rest of the country.

2.4   Luxury Is All About Motion

No, there is no typo in the title, luxury is all about motion, and not emotion as often believed. In this section we will discover how motion—the forgotten sixth sense—is driving the whole luxury industry. More than half of the trillion USD worldwide luxury market is spent on cars, yachts, private jets, and booze (Bain & Company, 2016). We will discover in this section the link between all these luxury items with the success stories of Moncler and Moutai.

Why do we spend so much money on expensive hobbies like golfing, clubbing, or sailing, and even pricey toys like drones? The answer, related to status-seeking, is in our joints, knees, tendons, skin, and ears, in the form of millions of tiny motion receptors.

2.4.1   The Sense of Motion

Often wrongly called proprioception or kinaesthesia, the sense of motion has a broader spectrum and encompasses the sensory signals received from our visual, vestibular, and proprioceptive receptors (the latter include also the kinaesthesic receptors).

This forgotten sixth sense—even denied by Aristotle—is a complex sense "that informs about the sensations of the deep organs and of the relationship between muscles and joints, generate afferent information that is crucial for the effective and safe performance of motor tasks" (Subasi, 2014). The tricky part is that all the received information need to concur. In other words, whether you will be a pilot, an Olympic champion, or just a clumsy "four-eyed turd" is directly linked to the sense of motion (The Silicon Valley, HBO, 2017).

Diseases, like neuropathy in diabetics, can lead to a degeneration of the motion sensors and to an impaired sense of motion. When elderly fall, unless you pushed them, it is mostly due to a less performant sense of motion, a consequence of a decrease in the number of vestibular receptors, which can give the illusion of leaning forward that is then overcompensated by a movement back leading them to the ground (Subasi, 2014). Actually, humans are made to be on the ground, so if you do not excel at any of this, you are the quite normal guy, others are just exceptional and therefore admired as champions and leaders.

2.4.1.1 Motion Receptors and Internal Mechanics
If you have some basics in car mechanics or your piloting license, what follows will be a piece of cake, if not, reading this book might also help you better understand some of your wheeled pets.

Motion sensors encompass the vestibular system, rods in our eyes, skin sensors, Golgi tendons, muscle spindles (Fig. 2.3).

The vestibular system acts like a gyro, with the semi-circular canals providing angular rotation information such as rolling, pitching, or yawing, and acting as an accelerometer, with the saccule providing gravity information, and the utricle providing linear acceleration information (or rate of change in velocity if you prefer) (Pilotfriend, 2016). Our internal gyro and accelerator work thanks to the little haircells swimming in the macular gel. Their movements are as many indications translated into electric signals sent to the brain.

You can also find gyro sensors and an accelerometer in car navigation systems, cameras, sports equipment (in golfing, for example), drones, and even in your smartphone. If we open your phone now (don't do this at home!), we would also find three tiny tiny accelerometers made of silicon, each giving information on one

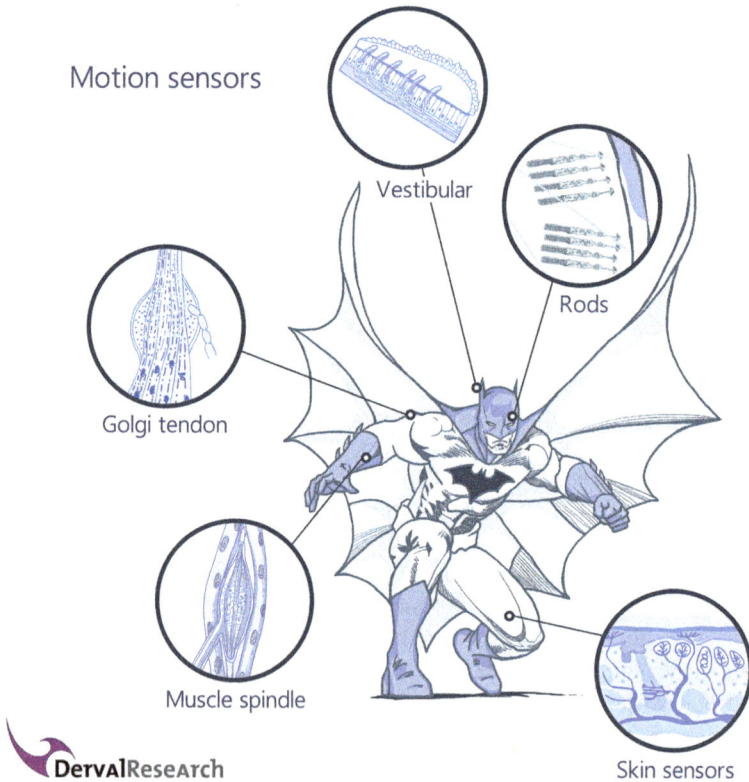

Fig. 2.3 Motion receptors (printed with DervalResearch permission, illustration by FangFang after Finger and Kane's batman character)

direction. They are a bit less big than the fiber-optic gyros you can find in racing cars and motorboats, or than the ring laser gyros used in airplanes and space shuttles, but accurate enough to tell when to rotate the display from portrait to landscape (Smith, 2012). They can also be found on a camera to correct the shake. These smart receptors take into account the Coriolis force—the one that makes wind turn to the right on the Northern hemisphere and to the left on the Southern hemisphere, and causes most of the plane accidents when pilots have the illusion to be turning because of the earth rotation when the plane is following its planned trajectory—for that particular case, pilot manuals even advise to better trust the navigation system than human senses (Pilotfriend, 2016). Hence the need for a luxury watch.

2.4.1.2 Neuro-muscular Control and Super Powers

If we look at the human body, muscles do much more than just contract and relax: they are our additional sets of gyros and accelerometers. Vibrations in our tendon, joints, and muscle proprioception receptors transmit movement and position information via the dorsal root nerve to our central nervous system, that then synchronizes with the vestibular and visual information to create dynamic sensory-motor maps based on the body motion and sending back information via the ventral roots nerve (Subasi, 2014). That is where probably all goes wrong for Asperger's like me, who have lower back pain, as it seems we cannot locate our own back (!), and are incapable of riding a bike, juggling, or tying shoe laces.

Each proprioceptor sends specific signals: muscle spindle provide critical information about muscle length and velocity, golgi organs about muscle contraction, and hair follicles endings on velocity and direction, so don't wax! Additional receptors that also give sensory input on touch and pressure serve our sense of motion when located in our joints: the Pacinian and Ruffini corpuscles (see Fig. 1.4). Pacinian receptors give information on direction and velocity and Ruffini receptors on pressure and angle (Derval, 2010a, b). They are our joints' accelerometer and gyro (Raja, Hoot, & Dougherty, 2011).

If you want to know where your neuro-muscular control system stands, you can perform the following test: have your index fingers point at and touch each other, then request the help of a friend or colleague (they will just wonder what type of luxury book you are reading, but it's ok) and ask him or her to move one of your arms while your eyes are closed. Once the arm has been moved away and while your eyes are still closed try to move the other arm only so that your index fingers point at and touch each other again. Some might be surprised how difficult it can be to locate one's own finger. In that case skiing and sailing is not recommended.

2.4.1.3 Motion Profiles

Unlike Batman, we do not all make the most of our motion skills. Batman is an unusual superhero, created by Bill Finger and Bob Kane (see Fig. 2.4), who does not have any super powers but instead uses all his resources—his intelligence, physical prowess, and martial arts skills, billion-dollar wealth, and associated technologies like the Batmobile—to counter villains.

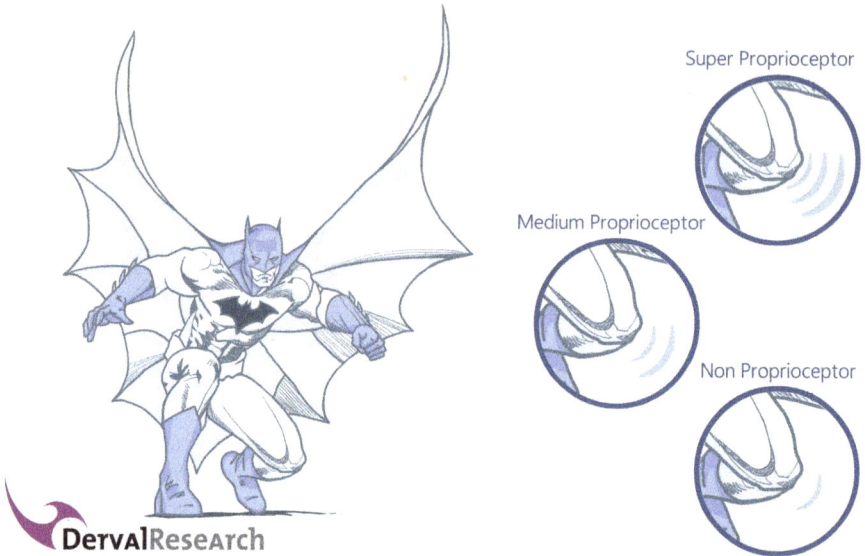

Fig. 2.4 Motion profiles (printed with DervalResearch permission, illustration by Fangfang)

Individuals display very various levels of motor skills, sense of balance, or aiming talents, so that we can segment people into three motion profiles (Table 2.2):

– Super-proprioceptors have fine motor skills and can perform all kinds of precision tasks like sailing, surgery, or aiming at a moving target. You might not want to play video games against them. They combine a great processing power with a speedy reaction time
– Medium-proprioceptors have comfortable motor skills and can perform various tasks like aiming, skiing, or driving with few hiccups. They have a good sense of balance and very reasonable reaction times but might be more subject to vection-induced motion sickness
– Non-proprioceptors have difficulties coordinating movements and might struggle a bit with tasks like lacing shoes, juggling, or even riding a bike. They have a poor reaction time. They can perform better in slower activities like some martial arts or fitness or in activities requiring less coordination. They would typically perform better at ping pong than at tennis as the service in the latter is too complicated for them

Reaction times are a good indicator of motor skills as they require coordination of brain and body. We used the reaction time ruler to measure with an accuracy of several milli-seconds the reaction time of luxury and non-luxury shoppers. The principle is that the measurer releases a ruler vertically and the subject has to catch it. The ruler indicates very precisely what the subject reaction time is.

Table 2.2 Preferences by motion profiles (printed with DervalResearch permission)

	Non-proprioceptor	Medium-proprioceptor	Super-proprioceptor
Motor skills	Has difficulties coordinating movements (lacing shoes, juggling, riding a bike)	Has comfortable motor skills and can perform most tasks	Has fine motorskills and can perform high precision tasks (complex gaming, surgery, sailing)
Aiming	Not very good at aiming	Good at aiming	Can hit a target even in movement
Balance	Feels more comfortable with both feet on the ground because of a poor sense of balance	Feels comfortable in most environments and has a good sense of balance. Can be subject to vection-induced motion sickness	Enjoys environments challenging a very good sense of balance
Luxury & leisure	Traveling, Opera	Ballet, Skiing	Off-piste skiing, surfing, sailing, waterpolo, racing
Reaction time in ms	Over 150 ms	Between 80 and 150 ms	Under 80 ms
Estimated population	25%	50%, more in Asia	25%, more men

Based on measurements and observations performed by DervalResearch on 1000 luxury shoppers in over 25 countries from April 2012 to March 2017

You can also try some of the online tests at https://faculty.washington.edu/chudler/java/redgreen.html, they are a bit less accurate, but fun. My best reaction time was 0.59. What about you?

We must admit that apart from highly paid pilots, heart surgeons, or professional F1 racers with an exceptional neuro-muscular control, machines created by men are much better at motion than their masters. And when talking with BMW drivers, many shared their excitement about driving and how they feel the car would confer them some super powers. So driving, piloting, skiing, sailing, quadding, or droning (not sure the terms exist yet, but who cares?), all these expensive activities are the price to pay for a better performance in motion.

Let's go back to our yacht story and see how motion profiles can help identify profitable markets. On a big yacht of 60 m, it is acoustically possible to limit the sound intensity to 35–40 dB, which would correspond to the sound of a car. On smaller yachts, it is trickier to harness the vibrations and while sunbathing on the deck you might be exposed to 70 dB sounds in the 50–100 Hz frequency—as if you would be following a Harley Davidson (J&A Enterprises, Inc., 2017).

For this reason, the Chinese, who are more likely to be medium-vibrators, do not buy superyachts but only megayachts! Like the 115-m megayacht powered by Lurssen bought by Chinese billionaire and real estate tycoon Samuel Tak Lee—his son took the opportunity to give a big party with David Guetta. Oh, but wait, Samuel is from Hong Kong, so it doesn't count. Hong Kong people are definitively more into boating as they live on an island. So selling a yacht to a mainland Chinese seems still a lost cause.

Especially since it has been proven that Chinese people are more sensitive, than for instance Afro-American and European-American people, to vection-induced motion sickness—that translates into unwanted gastric activity and elevated vasopressin (Stern, Hu, LeBlanc, & Koch, 1993; Stern et al., 1996). Vection is the illusion that you are moving based on visual cues like a yacht cruising or the landscape changing.

In the eyes, special receptors help detect movement: the rods. For instance when you are sitting in a pub with your fiancée and can't help checking other women passing by, you can just blame the rods: they are irresistibly attracted to movement. So the more you have of these guys, the more you are innocent. We will study rods and cones and vision in more details in Chaps. 3 and 4.

This vection-induced motion sickness—probably due to contradictory signals coming from our visual and motions senses—has been identified as causing gastric tachyarrhythmia and nausea also among Japanese, as measured by an electrogastrogram EGG (Imai, Kitakoji, & Sakita, 2006). Tachyarrhythmia is an irregular heart rhythm. The difficulty to adapt might therefore be because some people's heartbeat somehow aligns with the water waves or boat movement thus upsetting the overall system. I experience a similar phenomenon when listening to discordant music like certain types of Jazz or Bach as my heart beats tend to adapt to the music and become annoyingly irregular similarly to what happens in a vection-induced motion. Jazz hasn't made me puke yet, but I am very cautious! On the other hand, traveling in a straight line at regular speed relaxes me and makes me sleep (Levine, Chillas, Stern, & Knox, 2000). We will come back to rhythm and its influence on our perception in Chap. 5 about luxury and the perception of time.

So the reason why the yachting industry is not taking off in Asia is purely physiological. And it is the same reason Virtual Reality (VR) glasses are causing nausea in people subject to vection-induced sickness. Submarine types of luxury yachts with few and little windows, like the stunning Yacht A designed by Philippe Starck for a Russian millionaire, could be an option for Chinese UHNWIs, but they might still prefer a villa on the ground. Knowing the target consumers' motion profile is key in order to identify profitable markets and adapt the luxury and leisure experience.

2.4.2 Free-Riding and White Powder

Skiing is definitively a demanding proprioceptive activity and less subject to motion sickness issues. No wonder ski resorts are very attractive destinations. Alongside the trendy "skiing in Iran" opportunities, popular resorts are still Courchevel, Megève, Gstaadt, and of course St-Moritz.

If a majority of billionaires declare dedicating their spare time to charity and good causes, skiing, together with boating and wine and spirits are clearly in their top 20 activities (Wealth-X report, Billionaires Census 2015–2016). And these are even more popular among their many followers and other influencers. The wealthy or aspiring wealthy, following Batman's footsteps, tend to be fond of activities involving the sense of motion combined with equipment enhancing human abilities—understand that sports using human skills only, like football and Tour de France, are less attractive in spite of the various substances used.

By the way, declarative input (whether via surveys or focus groups) cannot be trusted: We made a funny experiment at a conference, where participants could text their answers on a digital wall. We asked the same questions—"what is your favorite drink and activity"—but the answers were "Coffee and video games" when participants' posts were identified and "booze and sex" when participants' posts were anonymous.

Courchevel 1850—this is not the year but the altitude, as there are several areas in Courchevel and the higher the more expensive—is packed with Michelin-starred restaurants, and luxury chalets. Megeve, created by the Baronne de Rotschild to compete with St-Moritz, offers seven five-star hotels and three Michelin-starred restaurants. Politicians and royalty are regulars at the Kolsters resort near Davos as it is the best place to ski off-piste and probably do some intense male-to-male competition while slaloming on the more dangerous pistes. For luxury shoppers, Gstaadt is the best-known, as major luxury brands like Chanel, Cartier, and Louis Vuitton have their outlets in the resort. Knowing well their target customers, Gstaadt also organizes every winter the Hublot Winter Polo Gold Cup—it is just brilliant how they managed to put so many sexy keywords together in just one name (Tisdall, Mawer, & Henderson, 2016).

St-Moritz stays the favorite skiing resort of achievers like Remo Ruffini, the Chairman of Moncler. In his spare time he also skies of course and when not, he may be on-board of his yacht (Forbes, 2017). The Moncler story is fascinating.

2.4.3 The Rise of Moncler

Moncler is a French and now also Italian brand created by two mountaineers in 1952 and they became famous for equipping the Olympic Team with quilted jackets. Remo Ruffini joined the declining brand as a creative director and bought it back in 2003, before introducing it on the stock market with an IPO in 2013. He introduced more colorful designs, fierce visual merchandising—putting in scene dinosaurs or sharks- and built strong distribution channels in strategic places like the resorts we just mentioned, but also in Miami or Shanghai (Fig. 2.5).

We had the chance to interview Remo Ruffini (2017) and when discussing the key ingredients for a successful luxury fashion design, Remo comes back on the motivations behind him investing in Moncler—which was by the way a great idea as the now French-Italian brand is ranked among the top 50 most valuable luxury brands in the world (Arienti, 2016): "When I decided to buy Moncler in 2003, the great heritage and history of the brand was absolutely unique, and that was the first thing I wanted to reignite. It was rare to find a brand with deep roots in tradition. The goal was to bring the brand into the future starting from its origins. Our aim was to roll out a "global down jacket" strategy all over the world. It was very important to communicate that Moncler is the jacket for all occasions. The brand's evolution has

Fig. 2.5 Moncler Window in Miami Bal Harbour (printed with Moncler permission)

become urban never renouncing the sporty spirit that is always present in Moncler's soul. This is what we really wanted since the beginning of this fantastic 'adventure', starting with a great product and an outstanding history. Nothing to archive but definitely to develop."

Moncler just opened a new store on the Kurfürstendamm, the main shopping street in Berlin, around Berlin Fashion week and gob smacked passersby and media with breathtaking visual merchandising, putting the outfit in scenes like no others (Visual Merchandising World, 2016).

When talking about the key ingredients for a successful Moncler window display, Remo reveals "I think that in a moment when the concept of rarity and uniqueness are constantly questioned, at Moncler we strongly underline that the ultimate reference for excellence is creativity. Creativity makes the difference, inspires admiration, evokes new scenarios and makes an identity stand out in a crowded, hyper-connected world. Creativity comes from research, investments, and fearlessness. These topics are at the base of all our activities, from Visual Merchandising to Events, and from Communication to Shows and Collaborations. Because if there is one element that determines excellence, out of the endless discussions on the actual relevance of fashion, it is its capability of surprising, while still maintaining its own DNA."

Remo Ruffini agreed also to share his definition of luxury: "I rather prefer the word elegance instead of luxury. In France there is a saying, from Sacha Guitry, that I often adopt: Le luxe est une affaire d'argent. L'élégance est une question d'éducation." This means "Luxury is a matter of money. Elegance is a question of education". Remo explains how since he was a child, his father Gianfranco has been a model to him: "his elegance and style really influenced me. Growing up I lived in the U.S. for a while, on the East Coast specifically. I was deeply attracted by the understated elegance and simplicity of the great families. Camelot, Hyannis—the beautiful Kennedys and their easy, classical flair, with their timeless elegance and classical touch. It was the beautiful American version of the important Italian families like the Agnelli's. I think that relaxed, effortless, chic elegance influenced me a lot. Even today, when I hear the word luxury, for me the immediate association is: something that expresses authentic and individual elegance beyond fashion. This is also how I describe my brand Moncler."

The brand, particularly appealing to super-proprioceptors, surely managed to develop and maintain a strong DNA and this is a great introduction to the brand codes framework we will study in Chap. 4.

2.4.4 Ganbei 干杯[1] with Moutai

Moutai is one of the top three spirits brands in the world and the uncontested leader in China with sorghum- and rice-based liquors. But what does booze have to do with the sense of motion? A lot, and it is linked to status-seeking and ranking. I used to be a party goer and noticed that men not only tend to make drinking a competition but most importantly to compare their performances once drunk. Who is still able to stand, who rolls under the table, who can still drive (do not do that at home, it is stupid and dangerous!), who can find the way back home, win a darts game, or touch the opposite elbow with the knee, standing on one leg only. They are benchmarking their motion skills in extreme conditions, a bit like free-riding in a snowstorm. Booze is therefore omnipresent at teams and also client gatherings in order to allow some kind of male ranking and identify who is number one and who is number two (Derval, 2016).

Moutai spirit became a hit when Chairman Mao Zedong presented it as a gift to American President Nixon in a groundbreaking trip to the U.S. in February 1972. The story doesn't tell which one of them won the motor skills benchmark under extreme (here Moutai) circumstances (Carew, 2011). This demonstrates the power of motion skills in luxury, and also the role of influencers as we will study in detail in Chap. 6.

Appreciating also the impact of travel retail in luxury sales, Moutai tried to name an airport in Huaihai after their brand by financing nearly 70% of its construction.

Making the most of the fascination for motion, the prestigious brand sponsors football and even named the Chinese Football Association Super League team after their brand.

---

[1]Means "Cheers" in Mandarin.

2.4.5   Motion Profiles: Business Applications

Now that you are aware of the critical role of the sense of motion in status-seeking and ranking and luxury, the endless business applications include:

- Designing any device or gear enhancing human performance, from golf accessories to spaceships
- Organizing gatherings around sophisticated activities, involving expertise and equipment, from driving to Michelin-star cooking
- Taking into account variations in motor skills when seizing the market for new technologies relying on the sense of motion, like Virtual Reality glasses as they would not be recommended for medium-proprioceptors subjects to vection-induced motion sickness
- Providing teams with opportunities to benchmark each other (racing, wine tasting)
- Offering luxury tools to measure and compare performances (wearables)

2.5   Turning Luxury Features into Customers' Benefits

Now it is clear why trying to sell yachts to most Asians might not work, but the Norwegian engineers still need help to understand what the sheikh really wants. What do luxury shoppers really want? How does one seize a market and evaluate a business opportunity? The Benefits Map and the Dyson and Stella McCartney cases will help shed some light on this mystery.

2.5.1   Stella McCartney: A Designer with an Opinion

Stella McCartney is a designer with an opinion. Pet and planet friendly, she went through much trouble and constraints in order to create a fashion brand exempt of animal products like leather and to respect animals and people all along the supply chain. And it is a success, as the brand, jointly owned by the luxury group Kering, is opening more and more outlets and strengthening its position in the fashionable activewear industry. Her collaboration with Adidas on colorful and printed sports-wear designed for movement has been very successful so far, maybe because Stella herself exercises almost every day: when not running, or cycling, she is doing yoga or horse-riding (Nilsson, 2016).

Stella McCartney loves horses. She designed an imposing 8000 Swarovski crystal piece horse showcased in her Las Vegas flagship store and the brand bought the rights for the painting "Horse Frightened by a Lion", by George Stubbs (Michael, 2013).

She loves horse-riding around the grounds of her country house, bought and patiently restored. The horse-theme follows her in the form of running horse patterns, or horse-printed trench coats and belts. Her love for horses and animals dictated the Stella McCartney major features, like the absence of animal-derived materials like leather and fur, but also wool whenever sheep are being molested (and it happens!). Because she is a vegetarian and animal activist, it is natural that Stella chose her friend Kenya Kinski-Jones, daughter of Nastassja Kinski and Quincy Jones, as one of the brand ambassadors for her fragrance POP. Kenya is an animal activist, but also a competitive horse-rider.

Stella doesn't mind being isolated and surrounded by trees rather than buildings. On her wedding list you could find trees instead of the usual cutlery: Gwyneth Paltrow offered copper beeches, Valentino and Tom Ford offered linden trees, to name a few. If her husband takes care of the garden, she is supervising the color palette. As it shows in her collections, she is more into pastel tones and has difficulties with red and yellow: some planted tulips supposed to be white grew yellow and she could not even look at them! (Bowles, 2010). The Derval Color Test®, in Chap. 3, will help you understand why.

Targeting empowered young medium to super-proprioceptors, Stella McCartney just designed a pair of white Adidas Ultra Boost shoes, with the Continental rubber soles (SNS, 2017).

The benefits of wearing Stella McCartney are double: identifying with the designer's lifestyle and adhering to her animal-friendly values in a very tangible way, not while wearing a Birkin bag.

With Stella McCartney designs taking into consideration pets, people, and the planet, the benefit for the followers are clear: being fashionable in a respectful way.

2.5.2 The Benefits Framework: Dyson and the Contrarotator Enigma

Talking about visionary entrepreneurs, Dyson revolutionized the quite boring consumer electronics category with bagless vacuum cleaners, fans without blades, and more recently hair dryers. The brand gained market shares in the upright vacuum cleaner market like in the U.K. and the U.S. as well as in the cylinder vacuum cleaner market like in Japan and Australia, and this in spite of retail price higher than the competition. In the U.S., Dyson has seduced 20% of the U.S. market, with retail prices three times higher than competitors. It was even question of a "Dyson-effect" in Australia, where consumers were also willing to pay a premium price for the vacuum cleaner. We can debate whether their products are luxury or not but to the least we can agree that they are premium.

Consumers are looking for "efficient, eco-friendly, ergonomic, and low-noise producing vacuum models". The centrifuge power used enables a more effective vacuum cleaning without any bag and without rejecting any air. For whom is this product a must have? Who are the target customers? The company decided to target allergic people or pet owners, or people who are very conscious of the necessity of having clean air in the room. For this type of customers, the Dyson is a must have. For other people, it is a totally useless product, unless they are looking for status.

Countries like Spain, Italy, China, and Japan were seduced by robotic cleaners. Roomba manufactured by iRobot has been the most popular robotic vacuum cleaner so far with millions of units sold. The government of South Korea shared its objective to become the leader in robotic vacuum cleaners with one robot per household and Samsung is clearly developing in this direction, challenged by Dyson also entering this market.

The distribution of vacuum cleaners has diversified and now you can bump into them in supermarkets and even grocery stores. The idea is to attract consumers who are not really thinking of replacing their vacuum cleaners. Some channels are indeed better suited for exploring innovations, like the gifting areas in department stores, TV shopping, or Facebook viral videos. Over time and in countries where Dyson is less popular, the brand is arranging a booth with its different products in department stores or at the check-out area in high-end grocery stores, even in waiting areas in airports, using the "Wait Marketing" approach, consisting of promoting products and services while consumers are available waiting, and more receptive as we will study in detail in Chap. 5.

An immediate success, Dyson's Airblade is available in 34 countries. Many airports, hotels, companies, and schools have adopted the innovative system. James Dyson, genius inventor and entrepreneur, perfectly understood the market for his Dyson Airblade: the market of hand drying. Who hasn't fought in public toilets with impossible-to-grab paper towels or hand dryers blowing air as strongly as an asthmatic ant would do. Looking like a toaster for fingers, Dyson's invention dries hands two to four times faster than another hand dryer, while using 83% less energy.

James Dyson also invented other awesome products. Like the fan without blades, generating cold and warm air. The Dyson Air Multiplier creates a low air pressure by multiplying the speed of the air by 50%. The brand entered a more fashionable market recently with the hair dryer based upon the same cyclone technology.

Dyson always follows the same three steps:

1. Does my product solve a problem?
2. Can it do it better than substitute products?
3. Is my product a must-have for certain types of customers?

Dyson always starts with users' frustrations. For instance, Dyson saw that the electric hand dryer would not work very well, would consume a lot of energy, and that paper is not environmentally friendly, which is an issue. The same for the vacuum cleaner that was getting stuck and didn't vacuum properly.

It took Dyson 5126 prototypes before being able to answer "yes" to these three questions and to launch the first version of his vacuum cleaner without the bag. Distributors didn't get the innovation. Or maybe they didn't want to give up the revenues generated by vacuum cleaners with bags. The product would take 14 years to hit the shelves for this reason: a big lesson of determination. If you have a great idea, never give up! Especially if people are mocking you. Dyson products clearly appeal to performance-driven luxury shoppers, who, like the Sheikh, want the best of the best. In the yacht case there is a twist, as the Sheikh also wants to show off in front of his peers and needs therefore to have exclusive yachting equipment nobody has yet: the Sheikh wants tailormade, R&D on demand (Derval, 2011).

Dyson commercial successes have been widely covered, but have you heard of the brand's biggest failure: the Contrarotator? Using the same cyclone technology used in the vacuum cleaner, the air multiplier, the hand- and newly launched hair dryer, using the same bold colors that made its success, using as usual a pricing policy at least 50% higher than competitors if not more, using innovation that led to real efficiency with two drums rotating in opposite directions allowing to divide the washing time by two, Dyson managed with the Contrarotator to design a failure. Launched in 2000, Dyson washing machine has been removed from the market 5 years later. Qualified "an enigma" by James Dyson himself, the whole Contrarotator situation is like part of the X-files: not fully explained. The official version is that the price was actually not high enough and the company was losing money on each sale. Objectively, even big Dyson fans have never heard of this particular product. Also, must-have products always find their place in the market, even if pricey, as the sales volumes tend to unlock production costs savings. So what was really wrong with this product? They used their usual secret ingredients (cyclone, bold color and design. technological innovation, and superiority with measurable improved efficiency, higher prices than competitors putting them in the luxury category). I'll let you think of it for a couple of minutes.

The answer is that nobody cares about this innovation. To understand what happened you need to visualize the use of the products. First, picture an old lady vacuum cleaning. Her back is hurting. With a traditional vacuum cleaner, the whole process is painful and lengthy. With a Dyson the same old lady can complete the dreadful task in half the time, what a relief! Picture now an impatient business woman washing her hands in the washroom of a luxury hotel. In the past, she would have been drying her hands with paper towels that do not absorb water or turning her hands back and forth under a blow dryer as powerful as an asthmatic ant—very useful. Now she can put her hands in some kind of Dyson toaster and air is blowing in all directions on her hands that are dry in no time. Same for the Dyson Air Multiplier: before, people were using fans that just shook a bit of warm air or in worst case scenario ate your hair if it was a bit too long or too close. Now with Dyson, within a couple of seconds, the air is very cold. But take now the example of the Contrarotator washing machine. The old lady is back. She puts laundry in the machine. And then what? She stays in front of it, waiting for it to be done? Of course not, she walks away and does whatever old ladies do! So the fact the machine will be ready in 30 min instead of 60 min, seriously nobody cares. This is why the Contrarotator was a failure: the features were the same as usual but the benefits for the consumers were void (Bloomberg, 2012).

The tricky part is: how to translate luxury products' features into consumers' benefits. We can use the Benefits framework and the magic question "So what?" and apply it to the Dyson vacuum cleaner in three steps (Fig. 2.6):

1. Describing the main feature of the product or service in one sentence. Example: "uses a patented cyclone technology" for Dyson vacuum cleaner.
2. Based on this feature and with the help of the "So what?" question, identifying the benefits for the customer. Example: "vacuum cleaning goes twice as fast".
3. Asking the "So what?" question as many times as needed in order to find the ultimate customer benefit. Example: "Leaves more time for Netflix".

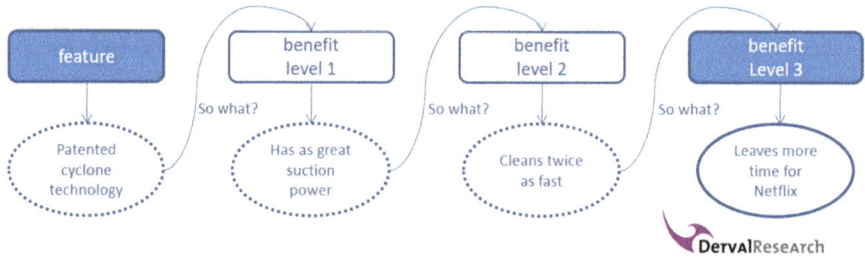

Fig. 2.6 Dyson vacuum cleaner benefits map (printed with DervalResearch permission)

Depending on the persona, the same product will have different benefits. In our yacht example, some will use the yacht size to show their power, while others will invite their friends or business partners for a sailing adventure to benchmark their motor skills, and others will just enjoy the sunset on the sea as we will see in Chap. 4 on our magnetic sense. In our performance-driven sheikh case, the key is to propose equipment other sheikhs do not have yet and that will improve the motor skills in terms of velocity or agility.

2.6   Take-Aways

Status-Seeking

– All rich people are not looking for status, but all people looking for status are trying to become or at least look rich, and are investing in luxury items and activities
– For some people, status-seeking is a physiological need, in the same way as eating or drinking. For them, status is a need and luxury items are a "must-have"
– Luxury shoppers are motivated by performance, power, or popularity. Understanding the three Luxury Shoppers' Drivers can help luxury brands identify profitable markets, and adapt their range and reach

The Sense of Motion

– Luxury is all about motion and in the race to success, motor skills and proprioception are key
– Luxury brands positioned on luxury leisure like skiing or horse-riding are a recipe for success

Benefits' Map

– Turning luxury products and services into shoppers' benefits helps clarify and quantify luxury business opportunities

References

Al-Maktoum, M. B. R., & Bishtawi, A. (2006). *My vision: Challenges in the race for excellence.* London: Motivate.

Arienti, P. (2016). *Global power of luxury goods 2016 report.* Deloitte. Retrieved from https://www2. deloitte.com/content/dam/Deloitte/ch/Documents/consumer-business/ch-en-cb-global-powers-of-luxury-goods-2016.pdf

Bain & Company. (2016, May 24). The global personal luxury goods market in 2016 will mirror last year's low single-digit real growth, even as geopolitical turmoil and luxury brands' emerging strategies reshuffle internal market dynamics. *Bain & Company.* Retrieved October 15, 2016, from http://www.bain.com/about/press/press-releases/spring-luxury-update-2016.aspx

Bloomberg. (2012, December, 18). Interview: James Dyson on killing the contrarotator – His educational failure. *Bloomberg.* Retrieved from https://www.bloomberg.com/news/articles/2012-12-18/james-dyson-on-killing-the-contrarotator-his-educative-failure

Bowles, H. (2010, October 22). *Stella's hideaway.* Vogue. Retrieved from http://www.vogue.com/865476/stellas-hideaway/

Carew, R. (2011, February 3). Moutai: A Chinese New Year 'Gan Bei'. *The Wall Street Journal.* Retrieved from http://blogs.wsj.com/scene/2011/02/03/moutai-a-chinese-new-year-gan-bei/

Dan, J. (2016, August 30). Advice of China's richest man goes viral. *China Daily.* Retrieved from http://m.chinadaily.com.cn/en/trending/2016-08/30/content_26640348.htm

Derval, D. (2010a). Hormonal fingerprint and sound perception: A segmentation model to understand and predict individuals' hearing patterns based on otoacoustic emissions, sensitivity to loudness, and prenatal exposure to hormones. In *30th International Congress of Audiology-ICA.*

Derval, D. (2010b). *The right sensory mix: Targeting consumer product development scientifically.* New York: Springer.

Derval, D. (2011). *Réussir son étude de marché en 5 jours.* Paris: Eyrolles.

Derval, D. (2016). *Luxury brand marketing 奢侈品品牌营销:创建·实施·案例.* Shanghai: Donghua University Publishing.

Derval, D., & Bremer, J. (2012). *Hormones, talent, and career: Unlock your Hormonal Quotient℞.* New York: Springer.

Forbes. (2017). *The world's billionaires.* Retrieved from https://www.forbes.com/billionaires/list/

Fujimoto, S., Miyake, T., & Yamahira, K. (2015). Latitudinal variation in male competitiveness and female choosiness in a fish: Are sexual selection pressures stronger at lower latitudes? *Evolutionary Biology, 42*(1), 75–87.

Griskevicius, V., & Kenrick, D. T. (2013). Fundamental motives: How evolutionary needs influence consumer behavior. *Journal of Consumer Psychology, 23*(3), 372–386.

Griskevicius, V., Tybur, J. M., Sundie, J. M., Cialdini, R. B., Miller, G. F., & Kenrick, D. T. (2007). Blatant benevolence and conspicuous consumption: When romantic motives elicit strategic costly signals. *Journal of Personality and Social Psychology, 93*(1), 85.

Griskevicius, V., Ackerman, J. M., Van den Bergh, B., & Li, Y. J. (2011). Fundamental motives and business decisions. In *Evolutionary psychology in the business sciences* (pp. 17–40). Berlin, Heidelberg: Springer.

Imai, K., Kitakoji, H., & Sakita, M. (2006). Gastric arrhythmia and nausea of motion sickness induced in healthy Japanese subjects viewing an optokinetic rotating drum. *The Journal of Physiological Sciences, 56*(5), 341–345.

J&A Enterprises, Inc. (2017). *Welcome to J&A Enterprises, Inc. noise and vibration control engineers.* Retrieved from http://www.jandaenterprises.com/

Kastanakis, M. N., & Balabanis, G. (2014). Explaining variation in conspicuous luxury consumption: An individual differences' perspective. *Journal of Business Research, 67*(10), 2147–2154.

Knight Frank Research. (2016). *The wealth report 2016.* Retrieved from http://content.knightfrank.com/research/83/documents/en/wealth-report-2016-3579.pdf

Levine, M. E., Chillas, J. C., Stern, R. M., & Knox, G. W. (2000). The effects of serotonin (5-HT3) receptor antagonists on gastric tachyarrhythmia and the symptoms of motion sickness. *Aviation, Space, and Environmental Medicine, 71*(11), 1111–1114.

Michael. (2013, October 17). Stella McCartney's lucky spot Swarovski horse chandelier. *Lightopia*. Retrieved from http://blog.lightopiaonline.com/lighting-videos/stella-mccartneys-lucky-spot-swarovski-horse-chandelier/attachment/stella-mccartneys-lucky-spot-swarovski-horse-chandelier/

Miller, T. J. (2017). *The Silicon Valley* [Television series]. HBO.

Nilsson, F. (2016, May 11). *Adidas by Stella McCartney*. Fitness on Toast. Retrieved from http://fitnessontoast.com/2016/11/05/interview-with-stella-mccartney-on-her-adidas-collection-for-ss17/

Palma, M. A., Ness, M. L., & Anderson, D. P. (2017). Fashionable food: A latent class analysis of social status in food purchases. *Applied Economics, 49*(3), 238–250.

Pilotfriend. (2016). *Pilotfriend general aviation portal – Pilot resources and aviation weather for general aviation*. Retrieved from http://pilotfriend.com/

Pozharliev, R., Verbeke, W. J., Van Strien, J. W., & Bagozzi, R. P. (2015). Merely being with you increases my attention to luxury products: Using EEG to understand consumers' emotional experience with luxury branded products. *Journal of Marketing Research, 52*(4), 546–558.

Raja, S., Hoot, M., & Dougherty, P. (2011). Anatomy and physiology of somatosensory and pain processing. In H. Benzon, S. N. Raja, S. E. Fishman, S. Liu, & S. P. Cohen (Eds.), *Essentials of pain medicine* (3rd ed., pp. 1–7). Philadelphia, PA: Elsevier-Saunders.

Reiss, S. (2004). Multifaceted nature of intrinsic motivation: The theory of 16 basic desires. *Review of General Psychology, 8*(3), 179.

Ruffini, R. (2017). *Interview by Diana Derval*. Amsterdam: DervalResearch.

Smith, D (2012, May 23). How does an accelerometer work in a smartphone? Bill hammack, the engineer guy, explains. *IBtimes*. Retrieved from http://www.ibtimes.com/how-does-accelerometer-work-smartphone-bill-hammack-engineer-guy-explains-full-text-699762

SNS. (2017, January 23). *Stella McCartney creates her own triple white Ultra Boost*. Kicks on Fire. Retrieved from https://www.kicksonfire.com/stella-mccartney-x-adidas-ultra-boost-triple-white-new-images/

Stern, R. M., Hu, S., LeBlanc, R., & Koch, K. L. (1993). Chinese hyper-susceptibility to vection-induced motion sickness. *Aviation, Space, and Environmental Medicine, 64*(9 Pt 1), 827–830.

Stern, R. M., Hu, S. E., Uijtdehaage, S. H. J., Muth, E. R., Xu, L. H., & Koch, K. L. (1996). Asian hypersusceptibility to motion sickness. *Human Heredity, 46*(1), 7–14.

Subasi, F. (2014). Posture, kinesis and proprioception. In *Proprioception: The forgotten sixth sense*. Foster City, CA: Defne Kaya.

Tisdall, N., Mawer, F., & Henderson, J. (2016, November 24). Top 10: Luxury hotels in the Caribbean. *The Telegraph*. Retrieved from http://www.telegraph.co.uk/travel/destinations/caribbean/articles/best-luxury-hotels-in-the-caribbean/

Visual Merchandising World. (2016). Moncler Berlin Fashion Week 2016. Retrieved from http://visual-merchandising-world.de/window-design/moncler-berlin-fashion-week-2016/

Wilson, C., & Kluger, J. (2013, October 21). America's mood map: An interactive guide to the United States of attitude. *Time*. Retrieved from http://time.com/7612/americas-mood-map-an-interactive-guide-to-the-united-states-of-attitude/

Finding the Right Positioning

3

"If you're horrible to me, I'm going to write a song about it, and you won't like it. That's how I operate."
Taylor Swift, record-breaking singer-songwriter (Diu, 2011)

In luxury, positioning is everything. This chapter shows, with the examples of Kate Spade, Y-3, and Porsche, how to identify blue-golden oceans, seize competitors, and grab new market opportunities.

3.1 Introduction

In this chapter, we check how to find the right positioning for a luxury brand and consider the following critical questions in Sect. 3.2 with the designer hand bag case:

- How to design "must-have" luxury items?
- How to find a profitable spot in the market?
- Are shoppers looking for premium brands also?
- Who are the real competitors?
- How to set up a pricing policy?
- How to grab new opportunities in the market?

We visit female-to-female competition backstage in Sect. 3.3, and unveil the huge role of entourage and social media in female so-called "relational aggression".

In Sect. 3.4, we realize with the Derval Color Test®, that we do not see the same things, and we evaluate the impact of colors on luxury products' appeal.

In Sect. 3.5, we see step by step how to identify the best positioning with the Nespresso, Porsche, and the Grand Optics cases.

Main learnings are listed in Sect. 3.6.

© Springer International Publishing AG, part of Springer Nature 2018
D. Derval, *Designing Luxury Brands*, Management for Professionals,
https://doi.org/10.1007/978-3-319-71557-5_3

3.2   How to Design "Must-Have" Luxury Items? The Designer Bag Case

Designer handbags is a fast-moving market: "one day you are in, and the next day you are out" (Heidi Klum, Project Runway, 2017). Hopping from Coach to Michael Kors or to Kate Spade, some luxury shoppers hate being mainstream. Let's see why designer handbags are a must-have luxury item (Schlossberg, 2015).

3.2.1   The Fashion Accessories Frenzy

Our new mission was to research women's thirst for fashion accessories, like designer bags. If we consider a clothing budget, up to half of it can easily be dedicated to heels and most importantly bags—we will see that those two often work together. Marie Claire magazine identified five clothing persona, and nicely called one of them "who do not bother about the middle part of their body" (Peng, 2016). One season celebrates Moschino bags, the next season the latest LV bag with painting masters like De Vinci or Rubens work-of-art prints, or the Gucci clutch. Meanwhile, some bags like the Prada Saffiano or the Hermès Birkin bag remain timeless classics. Good you can keep them as they are a bit pricey—from 1700 USD for the Prada to 11,000 USD for the Birkin bag.

The question being: What makes a designer bag so attractive and why would some women literally sell their mother to get one?

The best way to learn about designer bags was to observe women wearing them. Shadowing them like Stalklings (from the Trollhunters series on Netflix, written by Guillermo del Toro and animated by 88 Pictures in Mumbai for Dreamworks—I'm thinking of putting it as a pre-requisite for reading this book). We were for instance chatting with a Porsche dealer in Dubai when this powerful woman made her entrance. She was wearing a fitted black dress, showcasing her perfect silhouette, with a hint of red on her shoe soles. Her jewelry was subtle and very shiny. Her handbag was not only a very expensive designer bag, but also the latest model, and she was holding it with strength, hanging on her wrist. She was very interested in the Porsche Cayenne, maybe in red.

3.2.2   From Bags to Pockets

In ancient times, both men and women would carry little pouches attached to their waist. Wealthy ladies liked also to display these minaudières as a sign of status—it had a jewelry chain to which they could attach scissors, keys, and other useful items, looking a bit like the now popular Thomas Sabo chain with charms you can add to your handbag or wear as a bracelet. Then a major innovation considerably limited the use of bags: pockets. Added to men's pants or under women's wide gowns, they made pouches useless. At least until the next fashion revolution that happened in the eighteenth century when the hidden city of Pompei was unveiled and the Greek and

Roman tight outfits worn by the Pompei trend-setters became hip again. No way to hide pockets anymore and an opportunity to bring back the bags, mostly used for traveling first, then turned into cases or hand bags to accompany women during shopping, fine dining, or at work (Tassen Museum, 2016)

Today, men's purses represent also a nice 9 billion USD market (Karaian, 2013).

3.2.3 Which Bag Bitch Are You?

Back to our designer's bag and Porsche story. We let the vendor do his job and came back a bit later, curious about which car she finally bought. The vendor explained she was still hesitating with other brands. A couple of days later though she made up her mind and came back with her husband (she was a wealthy housewife) to confirm everything and buy the coveted car.

We soon unveiled at least four types of women regarding bags. The first type is carrying a practical bag "en bandoulière"—like a postman wearing the bag across the torso. She is more on the nerdy or tomboy side and seldomly wears high heels because … because you cannot walk with them of course! Then you have the woman who carries a more fashionable bag on the shoulder, nothing special here. Finally, you have the so-called "bag bitches". Beware, male readers, this is a warning: If you meet those women, run away or be prepared not to make any decision anymore for the rest of your life. The most terrifying one is the woman carrying her latest designer clutch—she spent all her or her sponsor's money on—hanging on her wrist. Wow, you need to be quite insensitive to pain to be able to carry anything on this body part. The idea is that carrying the bag on the shoulder would ruin her silhouette. She will be wearing killer heels and walking with a very natural ass-swinging move. For her, fashion means business and she has no mercy. The last type of woman is the one trying to be the ultimate bag bitch but too sensitive to pain: she can barely hold the clutch from the wrist and it always ends up higher on the forearm, where it hurts a little less. Also, even though she manages to wear high heels at the office during the day, she always keeps "real shoes" available in her big sports bag, to relieve her injured feet.

What urges these women to own the latest LV, Gucci, or Chloe bag? It is to seduce men? Probably not, as men couldn't care less about designer bags and fashion in general. Let's face it, the favorite outfit they enjoy on a woman is no outfit at all. So why do some women "need" the latest designer bag?

We will find the answer to this critical question for the fashion industry by having a closer look at the fierce world of female-to-female competition. As positioning is key for luxury brands, we will also have a look in the next section at who are a brands' competitors. Based on our designer bag case and our visit at the Porsche dealer in Dubai, would you be able to cite one brand competing with the Porsche Cayenne? I'll let you think a bit and we will review all the answers later in this chapter.

3.3   The Physiology of Female-to-Female Competition

Talking about this, handbags are a great biomarker (biological marker) of what type of woman you are facing and her level of competitiveness. Let's get real and ugly, and dive into the world of alpha females.

3.3.1   Relational Aggression

Each time you hear a woman assertively stating "I am a very competitive person", the question is how bitchy can she be? Recent feud between Taylor Swift, Nicki Minaj, and Katy Perry around the nomination of the best videos for the MTV Music Awards perfectly illustrated how passive-aggressive certain women can be when competing with each other (Lipshutz, 2015). In that matter, Swift publicly put down Perry in her video clip—using what psychology experts identified as being relational aggression trying to besmirch her image, and later she would also attack Niki Minaj, who was grumpy about awards always going to videos featuring skinny white women. Swift is clearly doing her best to establish her dominance over other celebrities and invited them all to her garden party for the 4th of July. But she was not putting down Katy Perry by coincidence: Katy happened to be the only female celebrity with still more twitter followers than her at that time (Meyers, 2015).

3.3.2   Becoming the Alpha Female

Similarly to the main antagonist of the fascinating FuerDai (new rich second generation Chinese) Canadian reality TV show "The Ultra Rich Asian Girls of Vancouver" about second generation filthy rich Chinese, featuring a group of 5 female "friends" spending their time enjoying a luxury lifestyle and back-stabbing each other (Thomas & Cornish, 2016). The dominant female—putting her friends/competitors down at every occasion—successfully ends the first season by marrying the wealthiest young man of the clique. Proof that all the bullying works.

We revealed, in Chap. 1 why men were into cars and into big, but why are women into bags and into shoes? Females are competing to get Mr Big and their weapon of choice is fashion and luxury and bitching aka "relational aggression".

Even among bees, female competition is great and is believed to promote migration from individuals to other nest sites (Janzen, 1966). In birds, female dotterels compete in arenas for access to males and the bright colored ones tend to have the advantage (Owens, Burke, & Thompson, 1994). Women's war can be more discreet and subtle as they will use relational aggression. Relational aggression might also be called back-stabbing or bad-mouthing and it is about harming someone by ruining or manipulating the relations to others—in other words her reputation (Dellasega & Yumei, 2006). Female baboons would mostly fight in order to secure a safe place in the middle of the troop, less accessible to predators (Ron, Henzi, & Motro, 1996). Fascinatingly, it was demonstrated that the female dominance rank

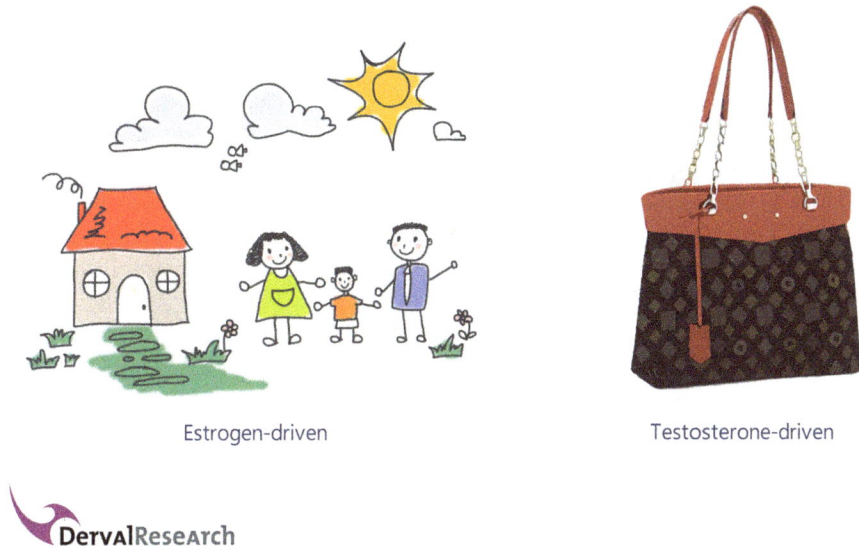

Estrogen-driven   Testosterone-driven

DervalResearch

Fig 3.1 The best gift ever by Hormonal Quotient® (printed with DervalResearch permission, illustration by FangFang)

and associated traits such as lower submission rate, contact avoidance, or contact aggression were directly related to prenatal androgen effects (PAE) or the influence of prenatal testosterone if you prefer (Howlett, Setchell, Hill, & Barton, 2015).

Women like men can be more influenced by testosterone or estrogen and it has an impact on their behavior and preferences, as we will see with the Hormonal Quotient® in Chap. 5.

In a previous investigation we did on gifting, we already observed bluffing disparities among women. When asked them to point at the nicest gift they had ever received in their life, some pointed at a child drawing (undoubtfully executed with more love than talent but, hey, who are we to judge? Plus it is sooooo cuuuuuute) while others showed us, with glowing eyes, their latest designer bag (Fig. 3.1). It doesn't necessarily mean that they like their children's artwork less but like that, top of the mind, a designer handbag is the ideal gift, especially when it is the latest one "à la mode" (on trend) (Derval & Bremer, 2012).

3.3.3 Ornaments, Ranking, and Luxury

Female territoriality has been highlighted among antelope marking wider territories, even outside of mating season (Roberts & Dunbar, 2000). When experiencing food stress, female grasshoppers tend to compete for the nurturant males (Castillo & Núñez-Farfán, 2008; Stange & Ronacher, 2012).

You know for sure that you are a competitive woman if you are straight and still check other women's booty in the street. In that case, you will likely spend budgets on ornaments intended to showcase your higher value, from jewelry to surgery. In ancient Turkmenistan for instance, women would wear as much gold and heavy jewelry as possible at their wedding, and until they gave birth. Then, they would loosen up a bit. The ornaments are supposed to show status and to bring good luck.

Nowadays also some women assess their value based on the content of their jewelry box. Think of Kim Kardashian double traumatized for being violently attacked and then robbed of all her jewelry in the rented villa behind Place de la Madeleine during fashion week in Paris. Considered like this, surgery might still be a more sustainable ornament.

One would think that all women aim for the highest rank in order to access the best male and more resources, but actually status comes at a cost: having to defend one's rank is very energy consuming and risky. Some individuals do not cope well with that type of stress and very happily leave the highest rank to the most competitive ones (Lindström, Hasselquist, & Wikelski, 2005). We will see later that this is often linked to the gender polymorphism and to the influence of prenatal hormones.

3.4   Luxury Is All Black-and-White (with a Pop of Color)

I had the pleasure to be a jury member at a fashion show and one of the talented designers was questioned on why all his designs were black. Perplexed, he replied "Because I like black". One jury member then went on a monologue on why this was not a satisfactory answer and he needed to better pitch his work yada yada. Based on physiological knowledge I will share with you in one second about color perception, I must say the designer's answer made total sense and you will discover why the world of fashion is all black and white, with a pop of color.

3.4.1   The Sense of Colors

Vision is a powerful sense and that aspect is an important factor in luxury. Let's see in detail how we perceive colors, patterns, and contrast.

3.4.1.1 Color Profiles: The Derval Color Test®
The Derval Color Test® was designed when investigating people's color preferences and was used, for instance, to help define Richemont's new luxury collections. Taken by over nine million people worldwide, and featured in many media and publications, including the NMSBA Neuromarketing Yearbook 2017, the test revealed huge variations in people's perception of colors.

Why do some people prefer black and some hate yellow? Remember, Stella McCartney's aversion for yellow flowers. Some individuals are more attracted to contrast while others are attracted to colors and both groups would benefit from an

Derval Color Test®

© DervalResearch – www.derval-research.com

Fig. 3.2   The Derval Color Test® (printed with DervalResearch permission)

adapted sensory experience. The objective of the initial research, since fine-tuned with the input of people around the world who took the test, was to study the impact of eye specialization on color perception among Chinese adults.

The Derval Color Test® and our findings on variation in color perception and preferences were presented at the 2nd Asian Sensory and Consumer Research Symposium in 2016 in Shanghai and even inspired the creation of glasses enhancing the vision of color nuances in Mikhail Kats Lab at the University of Wisconsin (Gundlach et al., 2017).

The original experimental group comprised 60 Chinese men and 140 Chinese women aged 25–45 years with no reported health disorders. Subjects reported their lower-order aberrations: nearsightedness or farsightedness. They were presented a color strip with 39 color nuances and indicated how many and which ones of the colors nuances they were able to see, as well as their color preferences in general.

Have a look at the Derval Color Test® (Fig. 3.2): How many color nuances do you count?

Among the 200 Chinese adults, some reported they were able to distinguish 27 color nuances and others 45. When broadening the number of subjects to various countries and profiles, the number of color nuances counted extended from 9 to over 45.

Based on their perception of color nuances, luxury shoppers can be split into four groups (Table 3.1.):

– Dichromats. They count less than 20 color nuances and tend to prefer black, beige, and blue. They see colors like dogs, mostly in beige and blue. Favorite luxury brands include Burberry and UGG
– Trichromats. They count between 20 and 38 color nuances and like all kinds of colors and love the assortment of colorful brands like Shanghai Tang or Miu Miu as well as more somber colors depending on their mood and style
– Tetrachromats. They count 39 color nuances and tend to avoid colors like yellow, perceived as irritating. They tend to prefer harmonious matching of colors with limited contrast and patterns. Brands like Balmain or Courrèges suit them well
– Monochromats. They count over 39 color nuances and tend to prefer black with a pop of color, contrast, or shiny. The information provided by their rods is the one ruling and their eye is attracted to and counts the separation stripes in between color

Table 3.1 Derval Color Test® results (printed with DervalResearch permission)

Derval Color Test® results	Dichromat	Trichromat	Tetrachromat	Monochromat
Color nuances counted	You counted less than 20 color nuances	You counted between 20 and 38 color nuances	You counted 39 color nuances	You counted more than 39 color nuances
Favorite colors	You like to wear black, beige, and blue	You enjoy various colors as you can appreciate them	You are irritated by yellow, so this color is nowhere to be found in your wardrobe	You love black and white with a pop of color
Estimated population	20% of the population	30% of the population	25% of the population	25% of the population
Photoreceptors	With a gapped distribution of cones, you are likely to have 2 types of cones	With an average distribution of cones, you are likely to have 3 types of cones	With an overlapping distribution of cones, you are likely to have 4 types of cones	With rods information overruling cones information, you are likely to have more rods than cones
Luxury Brands	Brands with basic colors and patterns like Burberry, or UGG	Many brands from Kate Spade to Hermès	Brands without bright colors or too much contrast like Balmain	Black and white brands with a pop of color like Chanel or Louboutin

Based on measurements and observations performed by DervalResearch on 1200 luxury and non-luxury shoppers in over 25 countries from April 2012 to March 2017

nuances rather than the color nuances themselves. They are perfectly comfortable with a Chanel or Louboutin assortment, mainly black and white with a pop of red

Subjects who counted less than 39 color nuances are not necessarily identified as colorblind given that colorblindness tests focus more on color than on color nuances confusion. Also, current tested occurrences of colorblindness include very different cases from people who have a missing type of cone to people who have cones with overlapping color ranges so that they cannot distinguish green from red. For that specific case, the Enchroma glasses solve the problem by filtering out yellow and helping people see green and red again! (Cross, 2016).

People perceive more or less color nuances depending on the number, range, and distribution of their color cones, and this perception acts as a predictor of an individual's favorite and irritating colors. These findings helped explain variations in color preferences and adapt the sensory experience—like the colors, shapes, contrast, and finish used in new luxury collections—to different types of people. We will see in Chap. 4 the special role of shiny.

Brands need to identify where their shoppers stand in order to make the right decisions, as we will see in the Grand Optics case later in this chapter.

3.4.1.2 Cones, Rods, and Melanopsin Receptors

70% of our sensory receptors would be located in our eyes, with around 120 million rods, 6 million cones, and the recently discovered melanopsin receptors (Marieb & Keller, 2017).

Light is a vibration too, like sound and touch, but of photons traveling at a high speed of 300,000 km/s and the colors composing the light are absorbed and refracted in different ways by objects and by our eyes, as we saw with the Derval Color Test®.

Cones are specialized in a color range, expressed in nanometers. S-cones detect short-wave colors like purple or blue, M-cones detect medium-wave colors like green, and L-cones detect long-wave colors like yellow or red.

Rods are specialized in perceiving contrast, movement, shapes, and night vision (Derval, 2010).

In addition to the rod and cones, our retina hosts a third receptor, in the form of retinal ganglion cells containing melanopsin (Fig. 3.3). Depending on the individuals, between 0.4 and 1.5% of the retinal ganglion cells contain melanopsin. These receptors feed our master biological clock—also called circadian clock—based on day and night indications. If, like me, your circadian clock is a bit messed up, you might feel depressed at times or have a tendency to overeat. On the other hand, you can also write books at night as you then have no notion of time! These receptors also monitor our pupil dilatation and enable some blind people to perceive

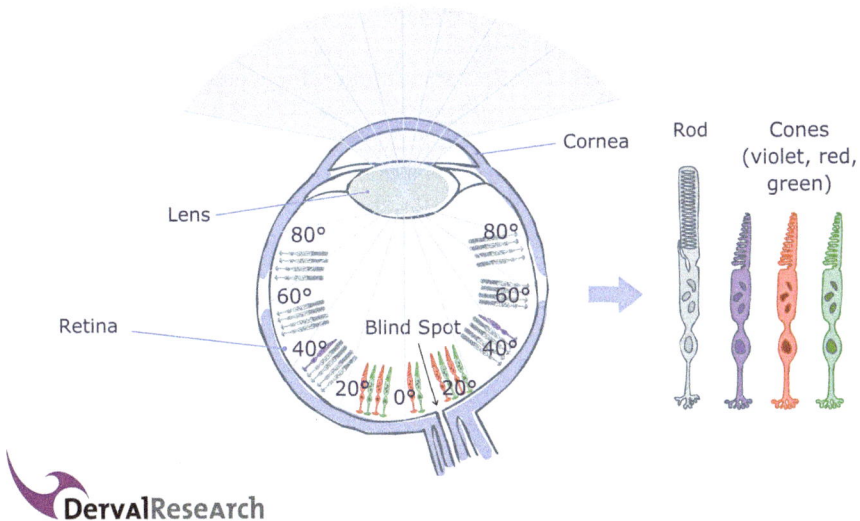

Fig. 3.3 Color receptors (printed with DervalResearch permission)

the presence of light even if they cannot see it. Similarly, bivalve molluscs do not see as such but have light receptors preventing them from being trapped.

Melanopsin cells are in contact with dopaminergic axons and form a kind of regular mosaic in the retina. Recent findings suggest that melanopsin receptors are not just sensitive to bright light, but to medium and long-wave length colors, whereas short-wave length colors are inhibiting them. Tetrachromats must have many M and L-cones and the associated melanopsin receptors and are therefore irritated not only by yellow but by bright light in general (Liao et al., 2016).

In some people, light and pain sensors are on top of it connected; which makes all pains like headache more painful when the light is on (Martenson et al., 2016). This photopollution—due to a too strong presence of artificial lights—can be painful for migraine sufferers and all people who have too active melanopsins in the brain, so that even some blind people can feel light aggravating their headache (Paddock, 2010) and it might also induce photosensitive epilepsy. Carl Zeiss designed cobalt blue lenses—the F133 blue has proven to be the most effective in that matter—protecting epileptic eyes against flickering but also red beams. The lenses are quite thick but block 80% of the light and especially red light. It is appreciated for cutting down the number of seizures (Belcastro & Striano, 2014).

We will see in Chap. 5 that melanopsin also plays a role in our perception of time.

3.4.1.3 Countries, Vision, and Color

When Kenzo introduced the best-selling fragrance Flower by Kenzo, their vision was to invent the poppy flower a scent. The elegant red flower produces poppy seeds and a special variety, the *Papaver somniferum*, produces opium. The designed perfume comes in a sober white packaging with a pop of red, as it features the flower and its long stalk, making it sensual (if you remember the stance discussion we had in Chap. 1). The scent is presented as a mix of mandarin, rose and violet, with a point of white musk—all rather subtle.

Flower by Kenzo was therefore the perfect product for Sephora China, combining the right scent (a subtle floral note) with the right packaging (white with a hint of red) to seduce local customers. We will come back to the sense of scent in Chap. 6.

Let's look into the human eye to explain the why. People can see colors with a wavelength higher than ultra-violet and lower than microwave radiations, in the 400–700 nanometers (nm) range. Each color has its wavelength: purple close to 400 and red close to 700 nm.

Favorite colors are often associated with local cultures and countries. We will see that it is only so because colors preferences are linked to our photoreceptors (cones, rods, and melanopsin), but also to our vision, and both are linked to our physiology.

F is the focal point of the eye, located in the retina, where all color waves meet after passing through the lens (Fig. 3.4). The exact location of the focal point varies between individuals:

- Farsighted people have a shorter eyeball and the F point is at the back the retina. A long-wavelength color like red will hit their retina at the focal point and will be perceived as relaxing

Color Perception

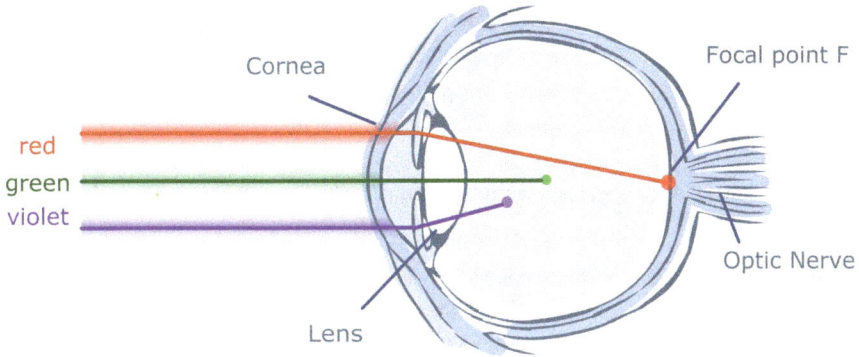

Fig. 3.4 Eye Focal point (printed with DervalResearch permission)

– Nearsighted people, so who have myopia, focus light in front of the retina. A short-wavelength color like blue will hit their retina at the focal point and will be perceived as relaxing

China has the biggest population of nearsighted people in the world with over 400 million individuals concerned. In the past 30 years, the proportion of nearsighted people in the USA increased also from 25 to 41%. To properly see red, near sighted people need to tense their ocular muscles, which can be more or less painful depending on the eyes' fitness, so that they might find a color like red exciting, and for some even irritating. Flower by Kenzo packaging is exciting to Chinese shoppers but not too much. Adapting locally is key for expanding a business.

Whether a brand changes city, province, country, or continent, it is important to first understand the reasons behind its success in its hometown before exporting the concept. In France for example, the best-selling perfume at Sephora would more be Dior J'Adore than Flower by Kenzo. Even though Chinese like gold, the scent is way too strong for most of them. People are different from one city to another and even within the same city, so understanding the features of a luxury product and mapping them with the local shoppers' preferences is key.

Environment and culture can be used to infer local people's physiology. Chinese people chose a red and gold flag as most of them are nearsighted and will find these colors exciting (Fig. 3.5). For Australia, it will be the contrary, as only 13% of the population have myopia, blue is very exciting for their flag (Derval 2010). And in France... now it becomes clear why it is so complicated to do business with these guys: they could not even agree on the color of their flag!

Fig. 3.5 Chinese, Australian, and French flags (printed with DervalResearch permission)

Brands need to adapt their sensory experience to every local market, and observing, but even better understanding, the neurophysiological mechanisms behind local cultures will help them be relevant.

3.4.2   When Street Meets Luxury: The Y-3 Case

Yohji Yamamoto is an interesting case to study as the acclaimed Japanese fashion designer made it without being French, Italian, or American. It is true that he had to come to Paris to be recognized in the haute-couture world.

Yohji Yamamoto is a leading and acclaimed fashion designer. With clients like Elton John, Placebo, or Adidas, he received many distinctions in Paris and Tokyo and is appreciated for his greater purpose: building a bridge between Chinese and Japanese people using the language of fashion. After some financial restructuring, the Yohji Yamamoto holding is now profitable again and the leading brands are the fashion brand Y's, the perfume brand Yohji Yamamoto Parfums, and the sportswear brand Y-3, a venture with Adidas. Y-3 brand is a high quality sportswear variation of Y's savoir-faire and includes apparel, accessories, and footwear for men and women, with 54 stores from Milan to Beijing. Each brand generates over $100 million revenues per year.

Y-3 brand codes stay true to Yohji Yamamoto's original values. The main color is black with a pop of bright colors like red, ideal for monochromats. According to Yamamoto "Black is modest and arrogant at the same time. Black is lazy and easy—but mysterious. But above all black says this: "I don't bother you—don't bother me.""

In terms of shapes, Yamamoto is a pioneer in asymmetric design and of course Adidas couldn't resist adding its famous stripes here and there. Yohji Yamamoto explained that initially he wanted to design men's outfits that also perfectly fit women.

Yamamoto's positioning would be anti-fast fashion. As the release of new products has accelerated in recent years, Yohji Yamamoto shares some wisdom about fast fashion and advertising: "Let me talk like an old man. Young people, be careful. Beautiful things are disappearing every day. Be careful. . . . You don't need to be shopping at fast fashion stores, especially young people. They are beautiful naturally, because they are young. So they should even wear simple jeans and a T-shirt. It's enough. Don't be too much fashionable.... The brand advertising is making you crazy. You don't need to be too sexy. You are sexy enough."

Even though Y-3 respects Yohji Yamamoto's original values, the style is definitively less high fashion but still avant-garde. A recent Spring Summer Y-3 collection featured for instance pictures found in social media under the topic "digital noise". The brand also celebrated its 10th anniversary on social media releasing a music compilation on SoundCloud.

This partnership serves Adidas brand architecture. As the company shared on its website: "We believe that our Group's multi-brand structure gives us an important competitive advantage. Through our brand architecture, we seamlessly cover the consumer segments we have defined, catering to more consumer needs, while at the same time keeping clarity of brand message and values. In each case, the positioning of Adidas, Reebok, and their respective sub-brands is based on their unique DNAs, their history, and their values".

In this footwear war opposing Adidas to Nike, Y-3 and fashion media exposure is playing an instrumental role. During last Mercedes Fashion Week, Adidas was able to attract, thanks to Yohji Yamamoto's designs, top NBA stars, and even Justin Bieber. In the same way, Yamamoto has access to the 15 million Facebook fans of Adidas: Y-3 is a truly win-win alliance.

Y-3 was the first brand to target the luxury sportswear market, followed by Stella McCartney. Adidas is trying to stay on top with innovative ventures—some remember iconic sneakers designed by Jeremy Scott, creative director of Moschino—and recently partnered with a Silicon Valley 3D printing start-up to bring custom shoes to the market. And as we saw previously with personas like the tuhao and brands like Moncler, comfortable and expensive is a profitable positioning.

3.4.3 Bold Colors and Affordable Luxury: The Kate Spade Case

Kate Spade is another great example on how to play with pops of colors to convert shoppers looking for affordable luxury.

Among the rising affordable luxury brands, Kate Spade is the fastest growing with now 300 stores worldwide. Many customers in China love the bold colors and reasonable prices. The US brand founded by award-winning designer Kate Spade in 1993 became popular for its convenient nylon bags but then got a bit forgotten. Bought by investors, it has been fully and successfully revamped (Derval, 2016).

The brand spent some time understanding which woman was the perfect persona for the Kate Spade DNA and developed a whole portfolio of items from bags to beddings that would sneak into her apartment and life (Fig. 3.6.): "[The Kate Spade woman] lives in a ten-floor walkup, but has champagne glasses. She doesn't take hours doing all of her holiday cards perfectly; she has a glitter party with her friends. Our brand promise is to help our girl live an interesting life, to live her life in color, in every sense of the word, it is not just about offering color, it is about living life to the fullest", shares CEO Leavitt.

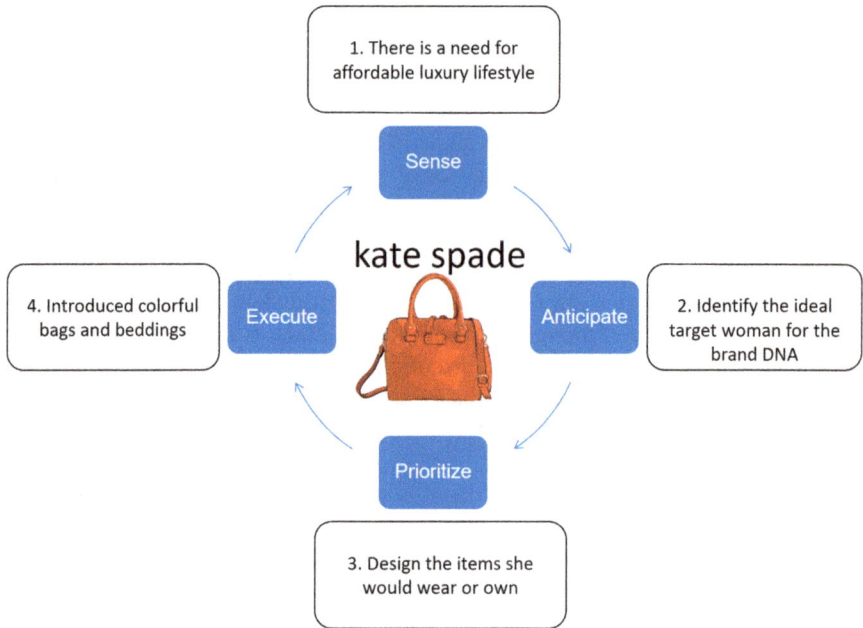

Fig. 3.6 Kate Spade strategy (printed with DervalResearch permission)

The portfolio includes bags, shoes, sunglasses, watches, and recently some additions like thermal coffee mugs, home décor, and bridal accessories.

Fudan University colleague Prof. James Yuann introduced me to the SAPE model—Sense, Anticipate, Prioritize, and Execute—and we applied it here (Fig. 3.6) to explain how Kate Spade was able to Sense there was a need for affordable luxury, Anticipate who the ideal target customer was, Prioritize by focusing on the items she would wear, and Execute the right strategy by introducing colorful items. Note that if handbags from all sides of the color spectrum are available, clothes are mostly black and white, making the overall assortment more appealing to monochromats. The brand is therefore very popular in cities like Shanghai where most fashionistas counted over 39 color nuances at the Derval Color Test®.

The brand developed a multichannel strategy with a presence in the best department stores as well as 300 branded stores and joint ventures, like in China with a regional headquarters in Hong Kong: "Where we think it is appropriate and impactful we have a strategy of forming joint ventures with local operating partners. We can then provide clarity around how we communicate to the consumer, be clear about the brand proposition, and take advantage of local partners and local operating expertise", confirms CEO Leavitt. This explains why the brand is planning its introduction in France only recently.

What stands out also is the success of fast fashion substitute brands including Zara—very appealing to dichromats for its beige and black tones and to monochromats for its various "shapes" of black—and H&M forcing haute-couture brands like Prada to open second line stores like Miu Miu, thinking they would rather copy themselves and make money out of it, rather than just letting others copy them.

The Chinese luxury market is huge and changing. So far, the luxury taste of Chinese consumers has benefited western fashion brands, mainly from France, and Italy. Affordable luxury brands like Michael Kors or Kate Spade are the new entrants. As it might not be the same consumers buying these different brands, understanding their typology is key. Even though the customer segmentation is now getting more complex, most luxury purchases are still driven by Chinese millionaires.

The three main fashion luxury buyers are indeed Chinese millionaires, eager to showcase their power and status with leading European luxury brands like Louis Vuitton or Gucci, business representatives using luxury goods for corporate gifting, even though recent government anti-corruption policies reduced that segment, and a growing Chinese middle-class with a more subtle taste and into popularity and performance rather than into power, offering new perspectives to fast fashion brands like Zara, and new entrant American brands like Michael Kors or Kate Spade, proposing affordable luxury with popular bright colors.

What is clear is that foreign brands add to the status. Now that new groups of Chinese can afford fashion and affordable luxury items, we see a shift in the market "from gold to silver" with a more refined and less logo-based taste in luxury.

3.4.4   Blending in or Standing Out? The Color Lenses Case

EyeQ is a strong online beauty player in the United Arab Emirates, proposing optical and beauty products such as contact and color lenses via the website www.lens2go.ae.

The site is a success and women can buy natural, enchanting, or trendy color lenses, depending if they want to blend in or to stand out. Founder of EyeQ and eCommerce guru Sajid Mahmood, who trained his optical skills with master Shawqi Ghanem himself, from Grand Optics (we will study at the end of this chapter), revealed during an interview we arranged at his brand new flagship store in the Reef Mall in Dubai that 99% of his customers are very young women. Usually they are the ones trying a lot: different make-ups, different fragrances, anything that can help them find their way, affirm their style, and for some, dramatically change their appearance. For having tried myself color lenses, it can truly and instantly metamorphose your face without surgery—very good as I am too sensitive for that. In terms of color lenses, if we consider the Emirati persona, or also the Filipino persona working in Dubai, more natural colors would be brown or green and stand out colors, grey. Kuwaiti brand Bella is doing extremely well with the local personas. Everything in their offering is consistent, from the name of the lenses—diamond or snow, very refreshing by 40 °C—to the associated pictures focusing on oriental eyes with a strong eye-liner and wonderbrow type of thick eyebrows. All that contributes, no doubt, to help Emirati online shoppers project themselves and pick the right product. For the Filipino customers, in addition to doing great deeds in order to equip needy people in their home country with vision glasses, EyeQ and lens2go partnered with Italian brand ADORE, proposing trendy styles. Recent campaigns helped reveal also that Korean beauties were more appealing to this persona than western beauties.

The brand is using Instagram to convey those messages to their target personas. Due to the local dress code, fashion accessories and eyes are the focal point in the seduction and power game. You can even spot eye-lash extension services in trendy Dubai malls!

Accessories and cosmetics, in that case color lenses, give women the power to change their appearance and beat their competition. And this has no price. We will see in Chap. 5 that cosmetics and fashion are often used to mimic different gender polymorphisms.

3.4.5 Color Profiles: Business Applications

Female-to-female competition is a huge driver for many luxury related services such as:

- Ornaments from cosmetics to surgery, with of course jewelry, bags, and shoes
- Exclusive clubs, offers, and happenings recommended by the favorite celebrity or blogger such as private sales and discounts
- As most items featured in social media are a one-show, there is a great potential for luxury items rental
- Creating shopping experiences appealing to each color profile with the right colors, patterns, contrast, and shapes
- Many fashionistas are monochromats and black and white collections with a pop of color are often winning recipes

3.5   Spotting the Right Positioning

Finding the right positioning is the mandatory path towards the right luxury shoppers, as we will see with the Porsche, Nespresso, and the Grand Optics cases.

3.5.1   Identifying the Real Competitors: The Porsche Case

When creating or expanding a brand, managers tend to immediately look at competitors. It is often misleading because (a) competitors might be doing silly things, and most importantly (b) we might not be looking at the real competitors.

Let's go back to our Porsche Cayenne example in Sect. 3.2. This 5-seat luxury SUV model is very popular in Dubai. But who are the competitors? Ferrari? Land Rover? BMW? You are nowhere near the answer.

The Porsche Cayenne is often bought as a second car. It is not used to go from A to B, but to enjoy leisure time with friends or show off a little bit. So depending on the purpose, a yacht, a villa on the beach, or a nice piece of jewelry will do. We can use the benefit map from Chap. 2 to clarify our vision, then the real competitors will suddenly become obvious.

Competitors are not brands providing similar products but brands providing similar benefits.

I will let you recover a bit from this shocking news and we will see, with the Nespresso case, how the positioning map framework can help identify all these unexpected competitors and detect new opportunities in the market.

The 330 million luxury consumers located worldwide purchase luxury goods but also premium items including designer second lines, cosmetics, and small accessories. We will, of course, continue analyzing the success of both luxury and premium brands like Chanel or Swarovski in coming chapters. And we will see that there is a fine line between luxury and premium.

3.5.1.1 The Position Map Framework: The Nespresso Case

The Positioning Map will help us clarify the benefits of our offering compared to other products available in the market and regarding the customers' needs. This method is used by successful leading companies but has never been documented properly. Let's see with the Nespresso case how to approach that tricky subject.

A good Positioning Map will also help identify new opportunities and give directions on what type of product to develop. This is why it is extremely important to plan this activity as soon as possible when creating, revamping, or expanding a luxury brand.

3.5.1.2 Who Are the Real Competitors?

Nespresso sold millions of single-serve coffee machines and opened luxury shops in prestigious locations, often near a Louis Vuitton or a Galeries Lafayette outlet. In terms of taste, the positioning of a Nespresso cup is slightly more bitter than a Starbucks Via instant coffee cup because the coffee is more roasted. Competitors might also be a good cup of tea, or why not a can of Red Bull if you want to stay awake. Even if technically the Nespresso machine is similar to another single-serve device from brands like Senseo or Keurig, as it does not have the same strength and bitterness, it doesn't target the same personas. So all the companies who are making similar products are once again not necessarily competitors because often they do not talk to the same type of people. Remember our Porsche case.

So the real competitors might be doing other types of products.

This is even more true if Nespresso is bought as a gift, the positioning map would then be completely different (see Fig. 3.7).

For many luxury accessories like jewelry or watches, as well as small consumer electronics, like smartphones or tablets, gifting represents a huge part of the sales.

We could for instance distinguish useful gifts from silly or decorative ones. Also we can separate gifts that will provide the receiver with status, like a rose-gold iPhone, from regular gifts. Key to successful positioning is to understand what persona will buy the product or service and for which benefit or purpose.

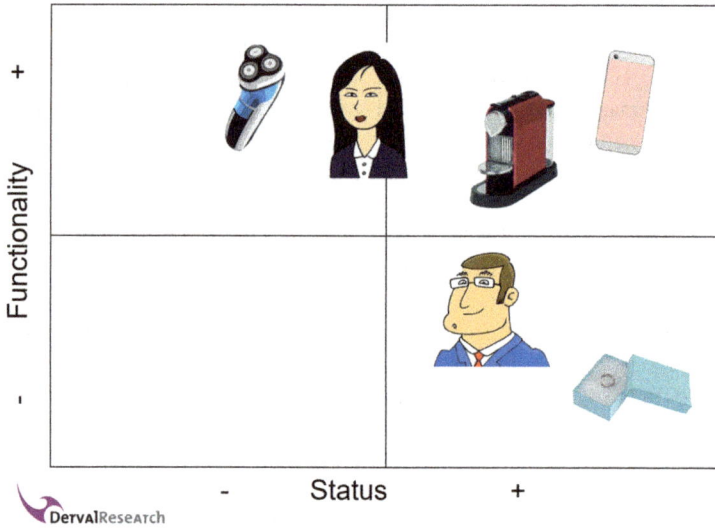

Fig. 3.7 Nespresso Positioning map (printed with DervalResearch permission)

So my advice is to focus on the persona. Describing the target persona and understanding their needs and motivations as well as the context of the purchase will help us identify the real competitors and benefits. Gary, for instance, is thinking of offering a trendy Nespresso Citiz to his wife for Valentine's Day, so they can enjoy a nice cup together. On the other hand, he thinks maybe a rose gold iPhone or a nice piece of Tiffany jewelry would be perceived as more romantic and give him more Xbox playing credits—as his "nuhanzi" (woman wearing the pants) controlling wife, is monitoring the access and usage of the gaming consoles in the house!

In another context, like Christmas for instance, Cindy is wondering whether she should offer her boyfriend a Nespresso but also thinks he might enjoy the latest Philips Shaver.

In the case of Philips shaving devices for men, 40% of the sales are due to gifting with a peak at Christmas and on Father's Day. Knowing that most gifters are indeed spouses and girlfriends, being appealing to women is key. And the brand even developed a campaign using an analogy with jewelry, explaining a Philips shaver is to the beloved one what a diamond is to a woman: "It's like you getting this little blue box"—the catch line is a direct reference to popular engagement ring maker Tiffany & Co—a success story we will analyze in Chap. 4.

It's a small world. So in the end, Nespresso is competing with a shaver, a piece of jewelry, and a phone. Probably not the competitors you would cite spontaneously.

The positioning map helps see new opportunities. For instance Nespresso, who did a successful partnership with Shanghai Tang releasing a limited edition of their coffee maker for the year of the dragon, could benefit from doing a co-branding activity with Tiffany for Valentine's Day. Starbucks and Swarovski joined forces recently proposing for Valentine's Day a lovely gift box with delicate chocolate roses and a pink heart-shaped Swarovski crystal.

3.5.1.3 Building the Perfect Positioning Map is an Iterative Process

Here some do's and don'ts for building a positioning map. First identify 2 to 3 personas who would buy the product and clarify the needs, motivations, and benefits: Who is buying the product for who? Is it for a gift or for a special occasion? What are the purchaser's drivers? Is he trying to seduce someone or just doing regular business?

Then identify for each persona 2 or 3 substitution products that can address the same benefits: What other products would the persona hesitate between when considering the purchase? Would he go in a specific shop near the office and check what is available there? Or have a look at Taobao or Amazon latest offers?

Make sure not to limit yourself to the same product category or certain product features. Focus instead on the benefits the product will procure. For instance a Nespresso machine can do coffee indeed but it will also provide the user with status as it is cool and trendy to own this machine. So any other item that can provide status—like a rose gold iPhone—is a possible substitution product.

While analyzing all these aspects, the labels to use on each dimension of the positioning map will automatically emerge. If we consider Nespresso as a gift then the positioning map is describing the gifting category and not the coffee makers' market. Two possible dimensions that would set gifts apart could be "functional"—as a gift might be more or less useful—and "status" as some gifts will make you want to share pictures on WeChat or Instagram and others less.

You can then place your 2 to 3 personas on the positioning map according to their sensitivity to the "functional" and "status" dimensions of the gift. Always try to identify specific terms for the two dimensions. Their purpose is to differentiate the personas and the substitution products. Terms like "quality", and "service", and "price" should be banned as by definition everybody wants the best quality and service for the best price. The real question is what does each persona mean by "best" product. For some, driven by performance, the ideal gift will be functional, for some, driven by power, it will give status, and for some it will be both or none. Price is basically what you are willing to pay for a product. So the more you see the benefits of a given product the more you are willing to pay for it. This is why it is not a valid dimension for our positioning map but a consequence of the product benefits that will be defined by two a bit more specific dimensions like in our case "functional" and "status".

Then, place near each persona the 2 to 3 substitution products they would consider for this gift purchase. This will give you an overview of the real competitors' landscape for a given product and help you refine your pricing, distribution, and promotion strategy.

We can see another illustration of the importance of positioning with the Grand Optics case.

3.5.2 Luxury Re-positioning: The Grand Optics Case in Dubai

The success story of Grand Optics shows us an extraordinary example of leadership and smart re-positioning from premium to super luxury.

Shawqi Ghanem studied electronics and started his career in a medical equipment company before moving to Dubai in order to manage two optics shops for his brother. In 1997, with a bank loan in his pocket, he decided to launch a new optical shop concept within a shopping mall. Grand Optics was born. Today leading a network of 70 optical shops located in the most coveted shopping malls in major cities like Dubai, Abu Dhabi, Alexandria, and Cairo and a team of 300 optics advisors, Shawqi reigns over the sunglasses business in the Middle East. And in this area of the world, sunglasses are a must-have for several reasons. First, the sun hits very hard and owning luxury sunglasses is actually not a luxury but a warranty of quality, as it means better-protected eyes. Not to mention that sand erodes the glasses so that they need to be renewed at least twice a year.

For those who wear a traditional outfit, sunglasses are a very strong differentiator, at the same level as a watch or a handbag. This ideal conjunction explains the strong margin generated by sunglasses in this area of the world: over 50%!

Grand Optics differentiated from competitors from the start with:

– New trends. Shawqi has the gift to detect the new models, colors, and sunglasses brands that will be a hit and Grand Optics is the favorite shop of many Middle-Eastern and Bollywood stars
– Shop delivery. 100% of the product references can be delivered within an hour. While competitors have only 80% of their references in stock and display delivery time between 24 and 48 hours
– Customer experience. Grand Optics offers the whole package in terms of customer experience with sales people listening and following up with customers, no matter in which country the customers are. For Shawqi, "quality is not a destination. It is a journey."

Shawqi, also a recognized poet, multiplies successful concepts with Mini Optics for children and an even more stylish network of optical shops named after his daughter Darah. At the heart of Grand Optics success is Shawqi's ability to identify the partners and location with the highest potential for his shops. This explains how the brand could adapt locally so brilliantly. Within a glance, this entrepreneur extraordinaire knows if a shopping mall will be successful or not. And he agreed to share some of his secrets with us!

Shawqi Ghanem agreed to share the six key indicators he takes into account when considering opening a new shop in a mall: "the profile of the shopping mall manager, the driver of the shopping mall (fashion, supermarket), the size of the gallery, the profile of the visitors, the location within the shopping mall, and the size of the shop".

3.5.2.1 Shifting from Premium to Luxury

Shawqi decided to reposition the brand on a more luxurious level, starting with the opening of his new Grand Optics shop in the prestigious Mall of the Emirates in Dubai. "The change in the demographics that some areas of Dubai have witnessed with the growth of wealthy communities and the construction of a lot of 5-star hotels and luxury vacation resorts was a motive for us to turn our shops, located in such areas, into luxury shops. In those locations, we eliminated from our collections all mid-market sunglasses and optical frames and started the offer with premium fashion brands such as Emporio Armani and Miu Miu, growing up to brands such as Gucci, Chanel, Prada, Dolce & Gabbana, Dior, Fendi, and reaching jewelry brands, such as Bvlgari and Cartier. The collections on display also include some solid-gold frames. In addition, we have reviewed our optical lenses in order to offer only high-end categories with healthier and more sophisticated features. The number of models on display has been reduced by 40% and this gives a much better feeling of exclusivity. All this is complemented by a very well-trained staff providing the customers with an unprecedented level of service and personal care. We expect that the new format will lead to reducing the number of transactions and to increasing the overall turnover and profitability. This new format has now been implemented in two shops at Mall of the Emirates in Dubai and in the YAS mall in Abu Dhabi."

With this super luxury revamping, Shawqi was facing a dilemma: "Should we completely change the shop design or maintain some elements that connect with our regular format? The success of Grand Optics with a large base of satisfied customers was built over 17 years and made it difficult to answer".

3.5.2.2 Appealing to Luxury Shoppers' Senses

Shawqi Ghanem built a luxury eyewear empire in the Middle-East, with over 70 outlets from Dubai to Cairo thanks to a great understanding of the brand target personas, the right malls to be present in, and a flair for the trends that will mark the coming season. Let's see how Shawqi made three strategic decisions in order to move to the next level. The decisions are related to positioning, visual merchandising, and assortment, and were decided based on neurophysiological insights generated by DervalResearch.

To better understand our target Middle-Eastern personas, let's go back to the neurophysiology of luxury. You remember from Chap. 2, that those with more dopamine DRD2/DRD3 receptors and people more influenced by prenatal testosterone are more into status and use luxury ornaments to serve their urge to climb the social ladder.

Middle-East is a paradise for luxury eyewear due to the obvious sun, the sand (which causes erosion that encourages frequent replacement of the glasses), the need for status, and the traditional outfit concentrating the differentiation on accessories like eyewear, bags, or watches. Measurements conducted by DervalResearch, with the help of the Hormonal Quotient® segmentation tool based on the influence of prenatal hormones, revealed that the fashionable target personas, both men and women, were clearly very-testosterone driven and eager to buy super luxury products. We will see in Chap. 5 how to measure your very own Hormonal Quotient®.

Therefore, the brand was clearly not over the top enough in terms of luxury. Grand Optics operated a successful repositioning from luxury to super luxury, switching from premium brands like Ray-Ban to exclusive brands like Tiffany—even proposing golden frames and other products made of rare and precious materials.

Many brands wonder which color to use but the real question is shall they use colors at all? Putting Grand Optics brand codes under scrutiny—white to convey the medical care values and orange as the historical brand color—it appeared that a more contrasted and less colorful mix would appeal more to the target personas.

Using the Derval Color Test® showing the different color nuances, we found out that the target fashionable crowd was more into contrast than into actual colors. Typical Zara clients, as opposed to Uniqlo fans. The same pipe transports contrast information coming from our rod receptors and color information coming from our cone receptors so that people counting more than 39 color nuances are dominated by their rods—we called them monochromats in Sect. 3.4.1—tricked into counting the separation lines in between the color nuances rather than the color nuances themselves.

Grand Optics increased the contrast of its visual merchandising, accentuating the black and white—favorite Emirati colors you find in the local dress code—adding just a touch of orange for contrast, making the shops visually comfortable places for their target shoppers to hang out and to buy (Fig. 3.8).

Shawqi confirms: "We decided to maintain the white clinical color that implies professionalism and healthcare, we kept the warm welcoming and fresh orange

Fig 3.8 Grand Optics new design in the Mall of the Emirates, Dubai (printed with Grand Optics permission)

color, in some furniture elements but in smaller proportion, and increased the portions of royal wenge wood color, as both represent the core values of Grand Optics, but we eliminated all old format furniture, and replaced it with new luxury elements, and this especially with the shop front—the first point of contact with customers—that we designed in order to reflect this new high-end positioning. This way, customers see the new Grand Optics but we continue building on the heritage of our very well trusted brand" (Ghanem, 2015).

Assortment wise, Carl Zeiss, a specialist in corrective lenses, proposes tailor-made lenses that correct the vision exactly on the spots where eyes need it, rather than simply in the center of the retina. The superior vision correction is particularly beneficial for people with many chromatic aberrations on the retina such as the bumps and holes displayed by the retina on the right (Fig. 3.9).

Given that the price of these tailormade lenses is three times higher than traditional lenses, and that playing the status and luxury card is difficult in that case as lenses are not visible and hardly show-off material, DervalResearch made measurements combining Carl Zeiss i.scription technology with the Hormonal Quotient® (HQ) and discovered that people who have a greater need for tailor-made correction due to higher chromatic aberrations also have a more testosterone-driven HQ—like the man whose retina is displayed on the right—while estrogen-driven people—like the man whose retina is displayed on the left—would benefit less from such a solution as they have less chromatic aberrations. In the same way as women have different favorite-ever gifts, as we saw earlier in this chapter, men also present different neurophysiological traits depending on whether they are estrogen-driven or testosterone-driven (Derval, 2010).

Grand Optics moved away from its initial partner Essilor towards Carl Zeiss' new pricey lenses and made a hit in the Middle-East as the benefits for its testosterone-driven consumers were instant. These three, in appearance, risky and strategic moves paid off and strengthened even more Grand Optics' position as an eyewear leader and iconic Middle-Eastern brand, now also active in other profitable

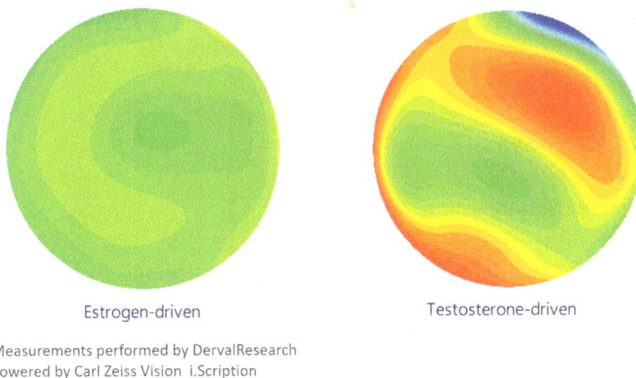

Estrogen-driven   Testosterone-driven

Measurements performed by DervalResearch
Powered by Carl Zeiss Vision i.Scription

Fig. 3.9 Retinal chromatic aberrations (left estrogen-driven man, right testosterone-driven man) (printed with DervalResearch permission)

businesses like children wear. Knowing target consumers' neurophysiological makeup revealed key in designing and revamping Shawqi's luxury brand.

A sharp understanding of consumers as well as retail customers' needs combined with an eye for style and latest trends enable Shawqi Ghanem to keep his brand up-to-date, be successful in every local market, and compete with large international brands.

The best way to predict which trends will be successful is to perfectly understand each persona, their product preferences, physiology, and motivations. For instance, a persona looking for performance might be attracted to novel shapes, materials, and colors, whereas a persona looking for power might enjoy golden frames. Key is to understand the motivation behind each fashion statement: why does this person wear this item? Once we have a clear answer it makes it very easy to predict which items the same person will pick in the new collection.

Many physiological factors will influence the purchasing decision: The body shape and face shape will influence the shape of the selected glasses, the vision will determine the favorite colors and finish (as we will see in more details in the following chapter), the sensitivity to texture and touch will influence the selected material and its weight—some people for instance cannot stand heavy frames as it is too painful, so even if they like to display status, the golden frames might be too heavy for them.

Luxury, super luxury, affordable luxury, and premium: it is not that easy to find the right positioning in the market. Also, competitors are not always the ones you would suspect, which makes it extra arduous for luxury brands. Some extremely successful brands like Kate Spade and Nespresso didn't hesitate to select a premium positioning. With the Grand Optics case, we saw how a powerful retail chain managed to reposition itself on the super luxury market. In all these cases, the positioning map can help identify the real competitors and discover new market opportunities.

Observing consumers—what they wear, how they behave—gives us clear clues on how they will behave in the future or when confronted to a new situation or a new product. We will see in coming chapter how to create and maintain distinct brand codes that can connect with your current and target luxury shoppers, and why shiny is so eye-pleasing to some.

3.6 Take-Aways

Female-to-Female Competition

- Some females are more competitive and use luxury and fashion to show their value and dissuade competition
- Female use more "relation aggressive" techniques making the most of social media to put competitors down
- Being in direct contact with customers is the only way to get to know them from the inside

Color Vision

- Depending on the number, range, and distribution of their color cones, rods, and melanopsin receptors, luxury shoppers will see color nuances, patterns, contrasts, and shapes in a different way and will develop distinct preferences
- Black-and-white with a pop of color is what rules the fashion industry, from fashion magazine Chief Editors to Key Opinion Leaders (KOLs) like bloggers, artists, and designers
- Depending on neurophysiological factors, such as the influence of prenatal hormones, luxury shoppers will see luxury products and services in a very different way

Positioning Map

- Finding the right positioning is key in luxury. It will confirm the personas and define the competitors
- Competitors are not brands providing similar products but brands providing similar benefits
- Competitors are more widely any brand tapping into the budget you covet
- Up-selling and cross-selling new products and services to existing customers first is way more cost effective
- Brands can be equally successful with a premium or with a super luxury positioning, as long as it stays consistent with the target personas and the brand codes

References

Belcastro, V., & Striano, P. (2014). Self-induction seizures in sunflower epilepsy: A video-EEG report. *Epileptic Disorders, 16*(1), 93–95.

Castillo, R. C., & Núñez-Farfán, J. (2008). The evolution of sexual size dimorphism: The interplay between natural and sexual selection. *Journal of Orthoptera Research, 17*(2), 197–200.

Cross, R. (2016, June 27). How EnChroma's glasses correct color-blindness. *MIT Technology Review*. Retrieved October 15, 2016, from https://www.technologyreview.com/s/601782/how-enchromas-glasses-correct-color-blindness/

Dellasega, C., & Yumei, D. (2006). *Relational aggression*. Retrieved January 10, 2016, from http://www.relationalaggression.net

Derval, D. (2010). *The right sensory mix: Targeting consumer product development scientifically*. New York: Springer.

Derval, D. (2016). *Luxury brand marketing* 奢侈品品牌营销:创建·实施·案例. Shanghai: Donghua University Publishing.

Derval, D., & Bremer, J. (2012). *Hormones, talent, and career: Unlock your Hormonal Quotient®*. New York: Springer.

Diu, N. (2011, April 3). Taylor Swift interview: 'I won't do sexy shoots'. *The Telegraph*. Retrieved from http://www.telegraph.co.uk/culture/music/8421110/Taylor-Swift-interview-I-wont-do-sexy-shoots.html

Ghanem, S. (2015). *Interview by Diana Derval*. Amsterdam: DervalResearch.

Gundlach, B. S., Shahsafi, A., Vershbow, G., Wan, C., Salman, J., Rokers, B., & Kats, M. A. (2017). Enhancement of human color vision by breaking the binocular redundancy. arXiv preprint arXiv:1703.04392.

Howlett, C., Setchell, J. M., Hill, R. A., & Barton, R. A. (2015). The 2D: 4D digit ratio and social behaviour in wild female chacma baboons (*Papio ursinus*) in relation to dominance, aggression, interest in infants, affiliation and heritability. *Behavioral Ecology and Sociobiology, 69*(1), 61–74.

Janzen, D. H. (1966). Coevolution of mutualism between ants and acacias in Central America. *Evolution, 20*(3), 249–275.

Karaian, J. (2013, October 10). Don't laugh at the "man purse"—it's now a $9 billion luxury business. *Quartz*. Retrieved October 15, 2016, from http://qz.com/133668/dont-laugh-at-the-man-purse-its-now-a-9-billion-luxury-business/

Klum, H. (2017). *Project Runway* [Television series]. HBO.

Liao, H. W., Ren, X., Peterson, B. B., Marshak, D. W., Yau, K. W., Gamlin, P. D., & Dacey, D. M. (2016). Melanopsin-expressing ganglion cells on macaque and human retinas form two morphologically distinct populations. *Journal of Comparative Neurology, 524*(14), 2845–2872.

Lindström, K. M., Hasselquist, D., & Wikelski, M. (2005). House sparrows (*Passer domesticus*) adjust their social status position to their physiological costs. *Hormones and Behavior, 48*(3), 311–320.

Lipshutz, L. (2015, July 23). Taylor Swift & Nicki Minaj's twitter argument: A full timeline of the disagreement. *Billboard*. Retrieved October 15, 2016, from http://www.billboard.com/articles/columns/pop-shop/6641794/taylor-swift-nicki-minaj-twitter-argument-timeline

Marieb, E. N., & Keller, S. M. (2017). *Essentials of human anatomy & physiology* (12th ed.). New York: Pearson.

Martenson, M. E., Halawa, O. I., Tonsfeldt, K. J., Maxwell, C. A., Hammack, N., Mist, S. D., & Heinricher, M. M. (2016). A possible neural mechanism for photosensitivity in chronic pain. *Pain, 157*(4), 868–878.

Meyers, S. (2015, July 23). Why female celebrities get stuck in public feuds. *Psychology Today*. Retrieved from https://www.psychologytoday.com/blog/insight-is-2020/201507/why-female-celebrities-get-stuck-in-public-feuds

Owens, I. P., Burke, T., & Thompson, D. B. (1994). Extraordinary sex roles in the Eurasian dotterel: Female mating arenas, female-female competition, and female mate choice. *The American Naturalist, 144*(1), 76–100.

Paddock, C. (2010, January 11). Light hurts even blind migraine sufferers, study reveals how. *Medical News Today*. Retrieved October 15, 2016, from https://www.medicalnewstoday.com/articles/175561.php

Peng, C. (2016, March 17). Fashion financials: 5 women break down how much money they actually spend on clothes. *Marie Claire*. Retrieved October 15, 2016 from http://www.marieclaire.com/fashion/news/g3557/women-on-clothing-budgets/

Roberts, S. C., & Dunbar, R. I. (2000). Female territoriality and the function of scent-marking in a monogamous antelope (*Oreotragus oreotragus*). *Behavioral Ecology and Sociobiology, 47*(6), 417–423.

Ron, T., Henzi, S. P., & Motro, U. (1996). Do female chacma baboons compete for a safe spatial position in a southern woodland habitat. *Behaviour, 133*(5), 475–490.

Schlossberg, M. (2015, August 5). Women have abandoned a longtime wardrobe staple—And that's terrifying news for Michael Kors, Coach, and Kate Spade. *Business Insider*. Retrieved October 15, 2016, from http://www.businessinsider.com/the-death-of-designer-handbags-2015-8?IR=T

Stange, N., & Ronacher, B. (2012). Grasshopper calling songs convey information about condition and health of males. *Journal of Comparative Physiology A, 198*(4), 309–318.

Tassen Museum. (2016). The history of bags and purses. *Tassen Museum Hendrikje Museum of Bags and Purses*. Retrieved October 15, 2016, from http://tassenmuseum.nl/en/knowledge-centre/history-of-bags-and-purses/

Thomas, A., & Cornish, D. (2016, June 7). *China's millionaire migration*. Special Broadcasting Service Corporation. Retrieved October 15, 2016, from https://www.sbs.com.au/news/dateline/story/chinas-millionaire-migration

Designing Luxury Brands

4

> *"The shiny red color of the soles has no function*
> *other than to identify to the public that they are mine."*
> Christian Louboutin, Iconic Shoe Designer (Collins, 2011)

For a mysterious reason, luxury is often all about glitter and shiny. We will see in this chapter what makes sparkle so appealing to luxury shoppers and how to smartly incorporate this knowledge in the design of luxury brands.

4.1 Introduction

In this chapter, we take a closer look at how to design luxury brands and review the following essential aspects in Sect. 4.2 with the diamond case:

- How to select a winning assortment?
- How to take into consideration local preferences?
- How to identify distinct brands codes?
- What makes a brand appealing and sustainable?
- Why do luxury shoppers love shiny?
- How to appeal to luxury shoppers senses?

We decode the mating selection process in Sect. 4.3, and find out that mating is closely linked to a taste for luxury and shiny.

In Sect. 4.4, we see through the examples of Tiffany & Co, Christofle, and L'Oréal that luxury is polarizing and reveal everything we know about the magnetic sense, discovered in humans only a decade ago.

© Springer International Publishing AG, part of Springer Nature 2018
D. Derval, *Designing Luxury Brands*, Management for Professionals,
https://doi.org/10.1007/978-3-319-71557-5_4

In Sect. 4.5, we see how to define unique and distinct brand codes with the Louboutin, Sofitel, and the Swarovski success stories.

Main learnings are grouped in Sect. 4.6 take-aways.

4.2   How to Select a Winning Assortment? The Diamond Case

I was guided into a tiny room—everything was black and shiny.

4.2.1   The VIP Room

The man offered me a drink, then grabbed a small piece of paper, wrote down some words, and then folded the paper and put it in a square envelope. He delicately put the envelope into a safety deposit box right behind him. He looked at me and smiled. I would later understand that the safety deposit box was in fact some kind of secret mailbox to send messages downstairs. After a couple of minutes, indeed, the envelope came back in the box with the most unexpected answer inside: a diamond, worth a quarter million USD.

We were in the heart of Amsterdam, in a diamond factory. Every year over 300,000 visitors are coming from around the world to enjoy a complimentary tour of the narrow building, designed that way in order to maximize the daylight throughout the day, for the cutters and polishers making live demos. The founder had the vision that providing visitors with deeper knowledge about diamonds would entice them to buy.

Diamonds represent a market of 80 billion USD and most diamonds are polished in the Indian province of Gujarat packed with skilled polishers, so that it is very special to be able to see this process in Europe (Chitrakorn, 2016).

4.2.2   Diamonds' 4Cs

Diamonds are evaluated based on 4 main criteria, the 4 Cs: Carat, Color, Cut, and Clarity. The carat is the weighting unit used and corresponds to the weight of a carob bean—which are fascinatingly always 0.00702 oz heavy (0.200 g). For a diamond, the heavier, the bigger obviously. The color defines whether the diamond is white or more yellowish. Color diamonds (brown, purple, or pink), are rated separately and can also be expensive. The cut is strategic and gives the diamond its sparkle. Tiffany is famous for its rectangular "princess" cut. The clarity assesses the number, size, and location of stones' imperfections.

First extracted in mines, diamonds are today also conceived in laboratories. These man-made diamonds are often shinier than their counterparts as they have no imperfections, but we will see soon that imperfections might be the secret ingredient of diamonds.

4.2.3   An Assortment Challenge

All these parameters have to be taken into account when preparing the assortment in the factory shop. Keeping in mind that it has to appeal to the visitors. But here, the visitors are coming from all around the world and Americans, Japanese, and Chinese will have a different taste diamond-wise. So what do you think: which one of the Cs is the most important to each group of luxury shoppers?

While you are guessing, let's explore this fascination people have for diamonds—especially ladies on the verge of getting engaged—and find out why perception can vary between individuals. A closer look at the neurophysiology of shiny might shed some light on why exactly some luxury shoppers are so attracted to dazzling diamonds and sparkling cars.

4.3   The Laws of Attraction and Mate Selection

Attraction to shiny can be closely related to mating and dating—think of the famous little blue box (we will study the Tiffany case later in this chapter). Mate selection seems like a complex process involving all senses but is very down to earth in the end, favoring short-cuts like mate copying, and making the most of all available biomarkers.

4.3.1   Mate Copying

Mate copying is a widespread mating tactic: you let others do the recruitment for you (Place, Todd, Penke, & Asendorpf, 2010). A bit like the headhunter calling you with this great job when you just got hired by a new company—so annoying. It is a smart use of resources as you might not have all information to evaluate a potential partner but his current girlfriend surely does. In guppies, a tropical fish species also known as the millionfish, even females genetically programmed to be attracted to males with more orange color on the body, would suddenly be attracted to a less orange male if observing that one or two other females show interest in him (Dugatkin, 1998). In humans, women tend to do mate copying only if the current girlfriend is attractive (Waynforth, 2007). For that reason, celebrities are often in the spotlight for boyfriend stealing, think of Brad Pitt moving from Jennifer Anniston to Angelina Jolie or Ashton Kutcher from Demi Moore to Mila Kunis. You are looking for a wealthy man, who likes children, and can afford personnel to take care of them? Simply become the nanny and steal the guy. Exactly what happened to, to cite only a few, Gwen Stefani, lead singer of No Doubt and mum of three. Luckily star country singer Blake Shelton, her co-jury on the Voice US was also available for a steal and they started a new season of their lives together. No wonder that four out of ten people registered on Tinder are in a relationship and three out of ten are married. Married men even share online that openly mentioning their marital status gets them more likes—some definitive cases of mate copying here too (Rushton, 2015).

4.3.2 Mr. Big or Mr. Right?

Mating preferences can be general to a population and influence its evolution. Take these two groups of guppies. In one, males have a stronger orange coloration, due to the fact that females always prefer the very colored males and others get little by little extinct (Houde, 1988). In the city of Shanghai, you have the particularity to find many "nuhanzi"—which in Chinese literally means "female-male"—and these women are clearly wearing the pants. Unlike in Beijing for instance, which is more male dominated. And nuhanzi tend to be attracted to another local specificity: "xiaonanren"—we could translate by little man and that depicts a more nurturing male, who spends probably more time shopping and grooming than his wife, as she is busy with the business and the money. It is said that "xiaonanren" give their salary to their mum and, when married, to their wife. So "nuhanzi" and "xiaonanren" complement each other very well.

Some people tend to look for their clones instead, especially if they have a high esteem of themselves. Finding a partner having the same sense of humor is a guarantee of liking the same things, sharing common values, and communications preferences. A study on married couples highlighted that most of them had the same sense of humor—measured by the Humor Scale (Hahn & Campbell, 2016).

When selecting partners, women do not hesitate to check whether the men are having a balanced diet by sniffing their body odor. A skin spectrophotometry revealed that sweat based on fruits, vegetables, meats, eggs, and fat was "more appealing" than sweat based on carbohydrates.

For *Metriaclima zebra* female fishes, wealthy and healthy is not all—they also like comfy. An experiment demonstrated that attractive males had to have some level of food resources on their territory but that providing a comfortable habitat was a definite plus (Greenberg, Jordan, & Sorensen, 2016).

4.3.3 Healthy, Wealthy, and Shiny

The way we walk, run, or dance—and from Chap. 2, even the way we sail or ski—tells a lot about us, our fitness (Hanna, 1987) and emotions (Fink, Weege, Neave, Ried, & Do Lago, 2014, Weege, Barges, Pham, Shackelford, & Fink, 2015). Our body language plays a great role in mate selection. Of course we consider not just the movement or posture but also multiple signals related to others' appearance, outfit, voice, and smell (Grammer, Fink, Juette, Ronzal, & Thornhill, 2001). Movement is complex and can be decomposed into speed, swing, and rhythm (Cutting & Kozlowski, 1977).

Males particularly pay attention to women when they are walking and it is suggested that the continuous movement from left to right provides an accurate measure of a woman's hip-to-waist ratio, a critical criteria in mate selection (Singh, 2002). Thanks to cues including hip-to-shoulder ratio, lateral body sway, and elbow position, we are able to recognize our relatives' gait (Troje, Westhoff, & Lavrov, 2005). The ornaments seem to matter more than the apparel. Male fishes of the convict Cichlid species are for instance programmed to be more sensitive to long wavelength colored female ornaments—the fish equivalent of Louboutin shoes, as we will see later (Fisher, Recupero, Schrey, & Draud, 2015).

Watch your walk as apparently shoulder sway is assimilated to male and hip swing to female gait (Cutting & Kozlowski, 1977). Women are on average a bit better than men in deducting cues from body language (Walk & Homan, 1984), but both men and women can read the seven innate emotions on people's body and not only from the face expression. For instance, a lower head port would be perceived as conveying sadness (Roether, Omlor, Christensen, & Giese, 2009). Also, we are able to distinguish biological movements—understand movements generated by creatures—as opposed to random movement, as they reveal "actions, intentions, and emotions", and this from birth (Allison, Puce, & McCarthy, 2000). Male spiders court females thanks to abdominal sways (Clark & Morjan, 2001), and male birds, with everything from neck and beak movements to feather erection (Patricelli, Uy, & Borgia, 2003), while female fishes keep it simple and prefer the fastest swimming male (Rowland, 1999).

Women, when evaluating a male dancing, consider fluctuating asymmetries—understood as little deviations in movements' symmetry imputable to health issues, genetics, parasites, or coordination problems (Brown et al., 2005)—so if you are a nonproprioceptor, do not dance; I repeat: do not dance. Particularly attractive are found dance moves from men who are stronger, as a research demonstrated measuring 30 males' dance moves as well as their handgrip (Hugill, Fink, Neave, & Seydel, 2009). Handgrip, face structure, as well as body movement are good predictors of circulating levels of testosterone (Fink et al., 2014, Johnston 2001, Derval & Bremer 2012), and of prenatal testosterone (Manning & Taylor, 2001). Referring to the Zuckerman sensation-seeking scale, males scoring high on thrill and adventure-seeking and disinhibition seem particularly attractive to women (Hugill, Fink, Neave, Besson, & Bunse, 2011; Zuckerman, 2007). The movements of the upper body (neck, abdomen) seem to make the difference between "good" and "bad" dancers (Fink et al., 2014).

Also, a skin rich in carotenoid—a substance contained in fruits and vegetables and giving them their yellow, orange, and red color—make men look more attractive (Zuniga, Stevenson, Mahmut, & Stephen, 2017). Now I understand better actor Orlando Bloom's success with women—must be his subtle yellow-orange tint. In general, orange/golden individuals are preferred by females whenever possible (Jirotkul, 1999). Female guppies also like males that have a nice carotenoid color obtained by synthetizing algae. But they should not be too orange, and not orange enough is also an issue. Only males who can balance their production of carotenoid with the right amount of produced drosopterins are attractive (Mark, 2014). So if you get a spray tan, don't do too orange. Carotenoids present two decisive common points with gold: the color wavelength and the shine.

If we go back to our diamond case, the slightly tanned Romeo would need, in order to convince his Juliette, to propose with a maximum of carats in the US, colorful diamonds in Japan, and shiny diamonds in China—with both the right cut and clarity.

We will see in next section that the fascination for luxury and shiny is all about the magnetic sense.

4.4   Luxury Is Polarizing

When I first arrived in Shanghai, I was overwhelmed by the brightness in the streets. Of course, the city is sunny during the day, but even at night the many shops in the streets are illuminated with such strong daylight bulbs—probably the result of Philips Lighting lobbying, but not only. Given the high number of golden iPhones circulating and the fact that most women outfits are encrusted with diamond-like ornaments, I sensed a fascination for shiny and luxury, as well as the local ability to absorb so much brightness.

4.4.1   The Magnetic Sense

The magnetic sense is the ability to use magnetic fields for different purposes like orientation, navigation, and migration, but not only (Foley, Gegear, & Reppert, 2011).

4.4.1.1 Another Dimension of Light
Let's start by a quiz on luxury cars and insects. According to you, would an aquatic bug—the ones hanging around near lakes and other water ponds—prefer:

(A) A red Ferrari
(B) A yellow Lamborghini
(C) A white Tesla
(D) A black Porsche ?

Surprisingly, even bugs have their preferences in terms of luxury. I let you focus and put yourself in the skin of an aquatic bug, or call a friend.

The correct answer is (D) followed by (A). How come?

When reflecting on a black or a red shiny surface like the roof of a metal, painted luxury car, the polarization (or direction) of the light is high and horizontal making the object very attractive to aquatic bugs, probably confusing it with water. Zoologists had been for a long time wondering why so many aquatic bugs were swarming, laying eggs, when not crashing on the roof of red cars. So you know which color to avoid to keep your fancy car clean in the swamp (Kriska, Csabai, Boda, Malik, & Horváth, 2006).

On the other hand, aquatic insects would not get caught in a mirage—a water illusion—as even if the mirage does look shiny from a distance, it does not reflect light with a horizontal polarization as water would do (Horváth, Gál, & Wehner, 1997).

Light can be characterized by its polarization (or direction), color, and magnitude (or intensity, if you prefer). Light is an electromagnetic field that oscillates and the direction of the beams are the polarization.

Animals use their magnetic sense and the light polarization to navigate, find food and drinks, communicate with their peers, and get potential dates. We must all have a polarization signature and I suspect mine is fuzzy, or unpolarized.

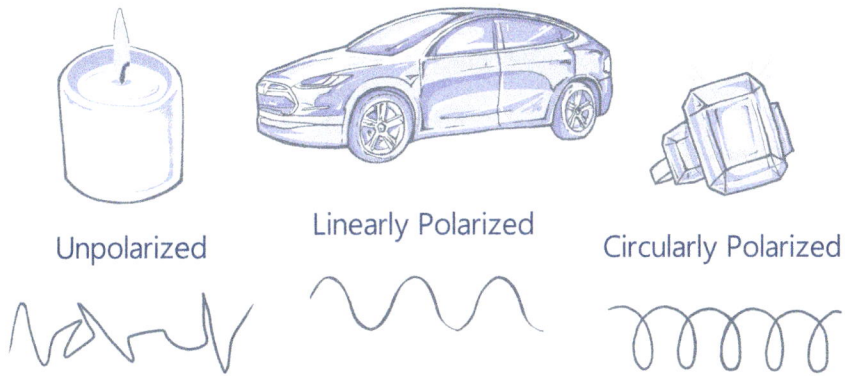

Fig. 4.1 Light polarization (printed with DervalResearch permission)

Light can be:

- Unpolarized. Like in the case of a candle, beaming in no particular direction
- Linearly polarized (vertically or horizontally), like in the case of the roof of a car or a sunset
- Circularly polarized. More rarely present in nature, like in the case of stars in the sky, a sunset on the sea, mantis shrimps, and diamonds (Fig. 4.1) (Marshall & Cronin, 2011, Roy 2006)

Animals and plants are sensitive to magnetic fields and some show polarotactic behaviors. The H_2O molecules in the water present in their body contain protons from hydrogen nuclei that become aligned in a magnetic field. We can think of the Earth, the Moon, or a Functional Magnetic Resonance Imaging (fMRI) machine. Novel investigation techniques like magnetic resonance works thanks to a big magnet—the fMRI machine—that attracts everything with a magnetic power of 3T (T for Tesla—I just realized Tesla was a measurement unit, before becoming a car!), which is 50,000 times more than the Earth's magnetic field. The fMRI allows us to spot blood flow movements as deoxygenated blood is paramagnetic while oxygenated blood is diamagnetic. Also, as the magnetic signal sent back by the hydrogen nuclei in the body varies depending on its surroundings, it gives a good indication whether it is near grey matter, white matter, or spinal fluids (University of Oxford, Nuffield Health Department of Clinical Neurosciences, 2016).

The sunlight and its reflection on water and shiny surfaces creates polarized light, and the angle or e-vector of polarization is critical. Water, for instance, presents a linear polarization. A perpendicular positioning of photoreceptors enables the measurement of the angle of polarized light (Wehner, 2001).

The magnetic sense has already helped solve many nature mysteries. This is how baby sea turtles, at least the ones that where not eaten up by birds, find their way to the sea without the need for any user manual.

In French, we used to say some people have a compass in the eye, which means they always find their way. Stunningly, migrating birds, but also wolves, monkeys, and sea otters, do have magnetoreceptors in the eye. A little flavoprotein called cryptochrome—from the Greek "hidden colors"—are indeed hidden in the photoreceptors and are sensitive to magnetic fields but only under blue light conditions, as it is activated by the S1 cones sensitive to ultra-violet (Gegear, Foley, Casselman, & Reppert, 2010). Pigeons even have extra micro-iron balls in their inner ear for better navigation (Nießner et al., 2016). So you have to imagine that the migrating bird is equipped with a windshield—his retina—where navigation data is displayed (maybe in the form of darker or brighter paths) so that it can comfortably auto-pilot towards its holiday destination. Similarly, honeybees use the sky light linear polarization to find food and also to communicate the location to their peers (Horváth, 2014).

In humans, melanopsin seems to play a magnetic receptor role—as we saw in Sect. 3.4.1 on vision, looking into melanopsin M1 and M2, knowing that the role of melanopsin M3–M5 also present in the eyes is not yet fully comprehended. Melanopsin is also involved in the circadian clock we will study in Chap. 5. This confirms the hypothesis of a chemical-based magnetic sense versus a for example mechanical or electric magnetic sense found in some fishes.

But wait, how does a compass work? A magnetic compass has a loose needle pointing to the North. It was invented in 206BC by the Chinese. In the eleventh century, they started using it to find their way, soon joined by Western and Persian explorers in the thirteenth century. For ship navigation, a gyrocompass was pre-ferred—the gyroscope has no secrets for you since Chap. 1—as it finds the true north rather than the magnetic north and is not affected by possible interferences from the metal junk you can find on the boat.

Some people are definitively more gifted for navigation and some are also much more sensitive to the magnetic field. I speak under the control of dog owners, but the same research finding magnetic receptors in dog eyes, mentions that dogs prefer to poop facing North or South. It would be interesting to trace if this is specific to certain canine breeds or individuals. Some of their owners also have their magnetic preferences. Regarding luxury shoppers, the best translation is *feng shui* (literally, "wind water"). As mentioned, the Chinese invented and first used the compass because they were preoccupied by their houses' auspicious layout and energy. Shall the bed face North or West? Are all the elements like water and earth well balanced? Many luxury hotels like the Sofitel SO in Bangkok take *feng shui* into consideration in their interior design. With the recent discovery of magnetic receptors, many unexplained facets of human and non-human animal behavior are being unleashed.

4.4.1.2 Turning Water into Gold: The Haidinger's Brush and the Dress

Before explaining how some people turn water into gold, we need to define shiny and learn more about gold nanoparticles.

Shiny is all about how objects reflect infrared: "shiny things reflect infrared energy and do not radiate well, non-shiny surfaces emit well and do not reflect as much" (Swirnow, 2014). For instance, a shiny leaf will reflect more infrared than a matte-leaf.

Gold is a special metal that both reflects (bounces back) and absorbs ("keeps") some of the visible (white) light that is shined onto its surface, as well as infrared, and it depends on the shape and size of the gold nanoparticles (Conover, 2017). Gold absorbs blue light and makes light look more yellow so that people who have more cones in this range of nanometers will find gold as painful to look at as yellow, if not more (UCSB, 2015).

During the Derval Color Test® field research, it was noticed that subjects who liked gold and shiny could distinguish fewer shades of yellow and were more likely to be subject to the Haidinger's Brush effect—and saw the famous dress gold instead of blue.

Our macula—an area located at the center of the retina—works like a radial analyzer filtering especially the blue light, with a peak at 460 nanometers. It seems that it is filtering by turning blue into gold, created a phenomenon called the Haidinger's Brush (Fig. 4.2). This phenomenon is best observed at the zenith of a blue sky sunrise or sunset and with the dominant eye (Pomozi, Horváth, & Wehner, 2001). When submerged by all this blue, a system in the retina is turning part of it into gold so that you only see the blue around some kind of golden bow-tie. Vikings, like the pigeons we discussed earlier, were using this Haidinger's brush to navigate, but I guess I still have to use a TomTom (Horváth, 2014).

Our sensitivity to polarized light is linked to the density of dichroic carotenoid pigments (them again!) located in the eyes' macula. The Haidinger's Brush is an

Fig. 4.2 Haidinger's Brush (printed with DervalResearch permission)

Haidinger's Brush

DervalResearch

optical illusion occurring when exposed to a polarized or even a circularly polarized light on a blue surface like the sunset or sunrise on the sea. Depending on the density of carotenoid in the macula that can be from 1 to 4, people seem to detect more or less well light polarization—from 23° to 87°—while some cannot perceive this phenomenon at all (Temple et al., 2015).

If you remember the "what color is that dress?" frenzy on social media, something like half of the people saw it gold then blue, half of the remaining ones seeing it always gold or always blue. It seems like a Haidinger's Brush effect is a good test to know if you have a performant in-built radial analyzer and also your short-wave-length receptors count. For me the dress was terribly blue, unlike my alchemist friends who turn everything they look at into gold.

A famous example of the application of light polarization comes from Polaroid. Edwin Lang became an instant success with Polaroid the day his daughter complained about not being able to see the photo immediately after it was taken. This inspired him to develop the first instant camera based on his knowledge on polarization. In a nutshell, Polaroid film is a plastic sheet that contains little crystals sensitive to magnetic fields and light polarization and will therefore create an instant "imprint" of what you aim them at. Already back in 1939 and before this patent, Polaroid brought groundbreaking innovation by projecting with Chrysler the first 3D movie to American audiences. 3D and polarized glasses for skiing were the firm's initial core business.

Polarized sunglasses contain a filter that removes all light going in a certain direction—often the annoying snow glare. It seems that people who are subject to the Haidinger's Brush effect have built-in polarized sunglasses and are therefore less sensitive to strongly polarized light as their eyes are filtering them naturally, helping them to just enjoy the sunset on the ocean or snow without the annoying glare.

4.4.1.3 Polarotactic Profiles

What if I tell you, fascination for diamonds and gold and shiny, and mate selection are all explained in one word? And this word is carotenoid.

Let's have a look at our friends in the animal kingdom. Many insects species, like bees, flies, or beetles are attracted to polarized lights. Natural light would diffuse brightness in irregular waves. Surfaces like a lake, a skiing piste, or a road will reflect light polarizing it horizontally, making it very appealing to flies for instance. The traps created to attract insects are based on this finding. A complex shape such as a diamond will emit a circular polarized light, appealing to beetles and Shanghainese women, to cite a few, based on the observation that many people and ethnic groups who elected sunny places to live, particularly enjoy shiny ornaments and other bling.

Fascinatingly, all the beloved luxury items and activities come together again linked by their reflective nature. You remember we talked about how women were attracted to males high on carotenoid and also how people tend to be attracted to similar people. Carotenoid is also a major component of the macular pigment present in our eyes which protects us against computer blue light and brightness in general, so that fit individuals can better cope with brighter objects and activities and might even be attracted to them. Search no more, this is why people spend their time checking their iPhones. It is not just because checking new messages increases dopamine levels out of excitement but because staring at a shiny surface is terribly eye pleasing for polarotactic people. In ancient times, people would just sit hypnotized in front of the TV, whereas today they have been upgraded to watching Netflix on their tablet.

The macular carotenoids can be broken down into lutein and zeaxanthin—their absence is for instance involved in age-related macular degeneration (Rapp, Maple, & Choi, 2000) and they can be found in leafy green vegetables like spinach or in kiwi, grapes, and orange pepper (Sommerburg, Keunen, Bird, & van Kuijk, 1998).

As we saw, not all beasts are attracted to horizontally polarized light: the scarab beetle for instance responds to circularly polarized light, assumingly as it helps him identify his peers, all emitting a circular polarization, which is rare in nature (Brady & Cummings, 2010). Apart from golden beetles, the mantis shrimp is sensitive to circularly polarized light. The mantis shrimp has even been observed sending polarized signals to communicate thanks to the carotenoid crystals contained in its antennas—a bit as if he would be waving his golden iPhone (Chiou, Place, Caldwell, Marshall, & Cronin, 2012).

Diamonds, especially under the blue sky, will refract light from multiple facets, creating a circular polarization. Gemstones in general will fluoresce—in other words reflect—under a black light. A black light emits ultraviolet (UV) light and is slightly purple. It is used to reveal gems of all kinds—as they each produce their very own reflection—authentic bank notes, as well as, in CSI episodes, to detect blood or other body fluids (Konnen, 1985).

It was assumed that humans' visible spectrum is between 400 and 720 nm and therefore they cannot perceive UV lights nor infrared (IR) (Derval, 2010). Which is a pity, as gold is particularly shiny and scattering around 750 nm, in the near infrared area. But in a recent experiment, scientists were surprised to be able to see a beam of near infrared light, and it was green. The explanation by Dr. Vladimir Kefalov of Washington University School of Medicine in St. Louis is that "if a pigment molecule in the retina is hit in rapid succession by a pair of photons that are 1000 nm long, those light particles will deliver the same amount of energy as a single hit from a 500-nm photon, which is well within the visible spectrum. That's how we are able to see it." (Sci-News, 2014)

So the infrared beam blinking twice was seen as a green beam. Knowing that gold is scattering at 750 nm, if it would do it twice, it could be seen at around 375 nm, which is blue. So don't throw away that blue stuff in your backyard—it might be scattering gold.

Individuals can, based on our observations and measurements, be split into three groups, depending on their attraction to polarized light and shiny objects (Fig. 4.3):

– Non-polarotactic individuals are somehow repelled by polarized light, like copepods (little crustaceans), and will prefer matte to shiny materials (Lerner &

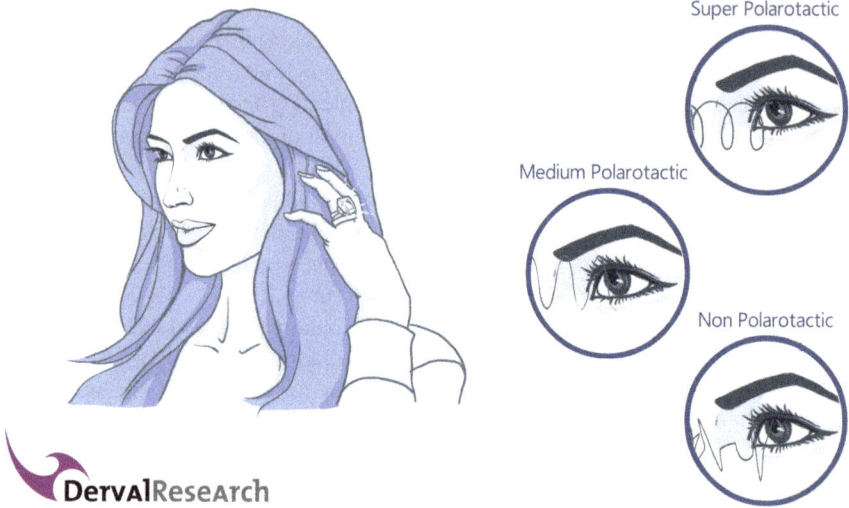

Fig 4.3 Polarotactic profiles (printed with DervalResearch permission)

Browman, 2016). They are not particularly attracted by sea and mountain activities nor fascinated by the sunset, stars in the sky, or diamonds in window displays. They prefer natural unpolarized daylight or why not, a candle light, and often wear sunglasses

- Medium-polarotactic individuals, like the orchid butterfly analyzing conjointly polarization and colors, enjoy shiny objects like cars, and activities involving horizontally polarized light like watching the sunset. They enjoy natural and artificial lights and wear sunglasses whenever needed
- Super-polarotactic individuals, like the mantis shrimp, are attracted to shiny objects like cars and shoes and bags (especially black and red), and polarized light whether natural like the sunset or artificial and wear sunglasses in very rare and specific occasions like driving or skiing. They are fascinated by circularly polarized light and could spend hours staring at the stars or at diamonds (Table 4.1)

If we go back to our diamond assortment challenge, Chinese buyers coming to the diamond factory are super-polarotactic and looking for the shiniest stone, whereas American shoppers are more medium-polarotactic and will also take into account the size of the stone, while Japanese shoppers are more non-polarotactic preferring colored stones—less shiny.

Table 4.1 Preferences by polarotactic profiles (printed with DervalResearch permission)

	Non-polarotactic	Medium-polarotactic	Super-polarotactic
Light	Prefers natural unpolarized light	Feels comfortable with natural and artificial light and likes polarized light	Is attracted to polarized light and especially circularly polarized light
Sunglasses	Often wears sunglasses	Wears sunglasses for special occasions like skiing	Hardly ever wears sunglasses as eyes are already filtering the polarized light
Surfaces	Matte surfaces	Shiny surfaces like cars or shoes	Sparkling surfaces like diamonds under a LED lamp, shiny surfaces (especially black and red)
Romance	A candle dinner	A sunset	The stars in the sky
Favorite hobbies	Indoors	Outdoors	Sea, mountain
Haidinger's Brush	No (saw the famous dress blue and black)	It depends (were able to see the famous dress in both ways)	Yes (saw the famous dress white and gold)
Estimated population	25%	50%	25% more in Asia
Luxury brands	Maxim's, Stella McCartney, Y-3	Christofle, Swarovski	Tiffany & Co, Ferrari

Based on measurements and observations conducted by DervalResearch on 1000 shoppers in 15 countries from April 2012 to February 2017. Findings presented at the 13th meeting of the European Society for Neuro-ophthalmology in Budapest and at the World Gem V conference in Chicago in September 2017

4.4.1.4 Photoreceptors Direction and Polarotacticity

Why is it so, that some people love staring at the sunset and at stars in the night sky, or can't get enough of admiring sports cars, diamonds and sports cars, while others could not care less?

With the help of retina models and cutting-edge optical equipment powered by Imagine Eyes, we decided to get at the bottom of it by having a closer look at the human eye, and especially at the luxury shopper's eye.

The device we used for our research is a non-invasive adaptive retinal imaging camera—worth 150,000 USD and designed by the French firm based in the technical hub in Orsay and emerging biotech city of Cambridge, MA. The rtx 1-e device is used by Imagine Eyes clients, surgeons and medical researchers, to display fine biological details never observed before, like the retinal mosaic—in order to prevent and follow photoreceptors degeneration that can occur for instance with diabetes-related neuropathy, or veins and arteries in the human brain.

Images of the retinal mosaic of the left eye of the subjects were taken at different eccentricities—different distances from the center of the retina, the fovea—showing

Super-Polarotactic   Medium-Polarotactic   Non-Polarotactic

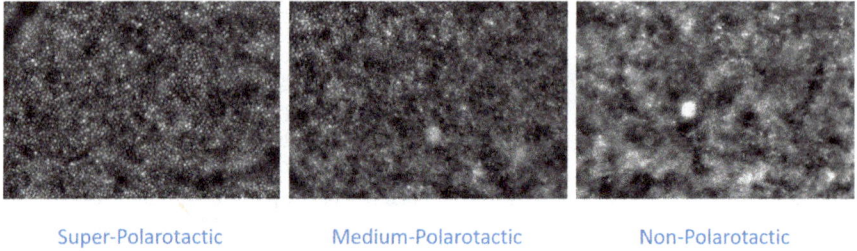

Fig. 4.4 Retinal Mosaic (printed with DervalResearch permission, imaging performed with Imagine Eyes' rtx 1-e retinal camera)

amazingly clearly the various photoreceptors. Nearsighted subjects tend to have a different distribution of the photoreceptors along the axon, and the concentration of receptors is known for being lower on the temporal side, hence the pictures in different locations. Exciting news is that the images were taken in-vivo—as opposed to on a dead body (yech!)—in a non-invasive procedure similar to a routine check-up at the ophthalmologist's.

The images that look like the surface of Mars or a busy metropolis from above (Fig. 4.4) are pictures of the retina, so clear that you can see the veins and the photoreceptors.

It can be observed that some photoreceptors are hyperreflective—they are brighter—while others are hyporeflective—they are darker. Also, photoreceptors are grouped to form some kind of sunflower-shaped patterns with one receptor in the center and on average 4–7 receptors around. It is not clearly understood yet with precision which ones are S-cones, M-cones, and L-cones, and rods, and if they are even the receptors themselves or just their container—as it is the case on the tongue where papillae are visible but the actual taste buds are hidden inside and arranged perpendicular to the tongue (Marieb & Keller, 2015).

Given the device is beaming a non-coherent and non-polarized near infrared light on the retina—mimicking the wavelength of a scattering piece of gold or a diamond sparkle due to nitrogen imperfections, if you remember—the hyperreactive receptors might be M-cones and L-cones as medium and long wavelength cones respond to infrared, or birefringent photoreceptors, just reflecting infrared. Imagine Eyes teams confirm that "thanks to their structure, the cone photoreceptors are in general hyperreflective compared to the inter-photoreceptor spacing, but the intensity of the reflected signal depends on several parameters, like the orientation of the cones, their cycle of regeneration (the signal strength seems to depend on the length of the cell), the presence of other structures above the photoreceptors that may absorb a part of the illuminating beam or project some shadows, the presence of diving capillaries". Interestingly, it is often judged that shiny eyes are healthy and as we saw earlier shiny means a better infrared reflectivity, partly due to a better processing of carotenoids.

Our small group of retinal models came from various backgrounds—male and female, Asian, African, and Caucasian, with shiny or matte preferences, 25–45 years old—and did not have major retinal issues. Subjects performed extremely well at the Derval Color Test®, seeing 39 color nuances for subject 3 (whose retinal mosaic is on the right (Fig. 4.4) and 38 color nuances for subject 1 (whose retinal mosaic is on the left), missing a yellow nuance, and for Subject 2 (whose retinal mosaic is in the middle) missing a purple nuance.

As the vision of colors depends on the number but also the distribution and range of the cones, it is difficult to draw conclusions for the count only, especially if the type of cone is not determined. Also Subject 3, who had a much lower counts than subject S2 and subject S1, is the one who was able to distinguish more color nuances. It could be that the count is more due to a variation in other factors.

For this reason, we tried to look more into the retinal mosaic pattern itself.

An interesting aspect, less researched, is the direction of the photoreceptors. Humans are not supposed to see circularly polarized light, unlike the mantis shrimp. The particularity observed in the underwater beast retinal mosaic (Cronin & Marshall, 2004), is that in a group of receptors, they all point in different directions, supposedly helping perceive circularly polarized light of his/her peers. It is therefore fascinating to know that in human also photoreceptors have a direction.

Looking at the retinal mosaic at the retinal eccentricity of 2°, subject S1 presents a very regular mosaic pattern with a hyperreflective dot every block, subject S2 presents a regular pattern with a small reflective spot here and there, and subject S3 presents a very irregular pattern with hyporeflective and hyperreflective patches. This could explain the lack of interest and even avoidance by subject S3 of everything shiny, as the patches of receptors pointing in the same direction might become overwhelming when light shines on them.

Subject S1, a passionate luxury shopper into cars and jewelry, loves shiny things, like cars, gold, diamonds, the sunset, skiing with the sun reflecting on the snow, and presents a typical super-polarotactic profile. Subject S2, an occasional luxury shopper into interior design and travel, is not excessively attracted to shiny and will sometimes choose matte surfaces and accessories over shiny ones, and presents a typical medium-polarotactic profile. The subject 3, an occasional luxury shopper into arts and accessories, is irritated by shiny objects and situations, would wear sunglasses more often, and always chooses the matte options, and presents a typical non-polarotactic profile.

Our attraction to shiny and luxury comes with variations in the retinal mosaic pattern and this opens the way to more exciting experiments. Maybe we could test this in real conditions, so feel free to send us your diamonds and sports cars—it is for science, of course.

4.4.2 L'Oréal Blond

If we go back to our dating scene, you might ask what about women—do they also send polarized signals using their golden iPhone? In many countries, they carry status symbols on their head: blond hair. The secret of L'Oréal might well be to use in their hair color the same substances—the shiny mica powder—used in car painting (Bengtsen & Kelly, 2016)—so avoid red and black, remember! Bleaching hair every 3 weeks can constitute quite a beauty budget but it's worth it. So if you spot a fake-blond, she is into status or she is trying to cover gray hair.

You might wonder whether colored hair would have an impact in the dating game. At least among birds like Gouldian Finches, it does. In this bird species, you find different morphs of males—we study this in more detail in next chapter—each with a different personality and head color. Females are particularly attracted to red-headed males, that are more aggressive than black-headed males. Funnily dying a red-headed male into a black-headed male completely fools the females (Roulin & Bize, 2007). Lesson learned: your L'Oréal budget might well be worth it.

The French brand originally called "la Société Française de Teintures Inoffensives pour Cheveux"—meaning "the French company for harmless hair dye"—is now heading a hair color market expected to reach 30 billion USD shortly so that even Silicon Valley investors are getting crazy about hair and funding Madison Color, for instance (Wohlsen, 2013).

Hair is a fascinating material that has its own orientation and can be defined by its a. shine, due to the reflection of light on the hair, b. chroma, due to the refraction of the light in the hair fiber and its reflection on the hair back surface, and c. diffused light, due to the depolarization of the light traveling through the hair and getting a particular tint. (Lechocinski & Breugnot, 2011). Under near infrared light, hair, fur, skin, and eyes fluoresce—which means they reflect the light. Because of the carotenoids and melanin (not to be confused with melanopsin) they contain, hair shines. It is unclear what humans are able to perceive under infrared light, but it would be a fascinating world, maybe similar to the one full of mantis shrimps that divers describe or the one full of shining stars that astronauts talk about.

Recently, researchers discovered that the nitrogen imperfections embedded in diamonds, that give them their unique color, were "alive" like the stars and called this phenomenon "time crystals". Similarly to crystals that grow in space and follow a certain pattern, "time crystals" grow in time following a certain pattern (Choi et al., 2017; Nayak, 2017). Therefore diamonds will look and react differently under the light depending on the moment in time. This explains why some of my dear friends can get lost staring at diamonds like we can get lost in someone's eyes, where 10% of the photoreceptors die and are replaced every day, creating a universe in movement (Chuang, Zhao, & Sung, 2007, Cronin & Marshall, 2011).

The *shine* is the same polarized light as the source, the *chroma* is a different polarized light as the source due to birefringence—a phenomenon by which a light ray becomes 2 light rays when passing through some materials like here the hair or the eye cones—and the *diffuse* is an unpolarized light. As you have light reflection on and in the hair, this gives hair quite some glow! Once again, red and black hair stand out and blond hair generates three times more oscillations than brown hair.

4.4.3 A Tiffany Blue Sky

Tiffany knows how to appeal to polarotactic shoppers. The brand uses the blue color to activate our magnetoreceptors and make its unique designs—little keys when others sell hearts—glow under rotating lights.

For instance, Tiffany's Korean flagship store uses the romantic "blue box" signature branding in the front display. The window display is literally surrounded by radiant blue light—you know the very light activating our magnetoreceptors. No wonder polarotactic women are attracted to the display like magnets. It is not to be excluded even that some women, under the Haidinger's Brush phenomenon, see half of the box blue and the other half gold—even better!

So Tiffany does not need to use gold to impress luxury shoppers as the gold is in the eyes of polarotactic shoppers. This is good news as the brand is originally famous for its sterling silver, like Christofle we will check right after. Most of their diamond rings are made of white gold though. Our favorite goldsmith, Hettie Bremer agreed to share some industry insights for us (Bremer 2017). Gold is extremely soft, so that you have to combine it with another metal. Nickel was used for this purpose in order to create white gold but it was soon discovered that many people were allergic to it and it has recently been replaced by the pricey platinum. But voilà—platinum is a bit greyish so most white gold rings are also polished with a layer of Rhodium, a super expensive and shiny material. This layer can fade and the super-polarotactic friends of mine regularly go back to the shop to get it shiny again.

Platinum, like silver, has a color spectrum closer to white, as opposed to gold more in the longer wavelengths, and is therefore a nearsighted friendlier option. It might also be an option for people whose skin tone is not particularly enhanced by gold and who move towards a rose gold or a white gold.

Tiffany is also famous for its diamond princess cut—the one Kim Kardashian is wearing. The more you have facets, the more the diamond is shiny as the light reflects on more sides. The princess cute looks quite impressive and is also more affordable as you can carve two from the same rough stones without wasting the edges, cut off in rounded diamonds. This shape is also likely to concentrate the light polarization in certain directions, increasing the intensity of the shine. Do not forget to send us your diamonds to help further research in that field!

Another optical effect you might want to know about is the Boehm's Brush, occuring when a polarized light rotates on a dark background—and creates a light scattering (Horváth, 2014). Jewelry leaders like Tiffany and Swarovski, discussed later, understood right away the power of polarized light.

4.4.4 Silver Ever: The Christofle Case

Christofle has been providing silver cutlery to kings and queens since 1830, specializing in fine silverware. The French brand proposes a luxurious assortment of tableware and home decoration. The brand became famous by being the tableware used in trendy places like Maxim's restaurant in Paris, where celebrities and

Fig. 4.5 Christofle shop in Shanghai Plaza luxury mall (printed with Christofle permission)

powerful people could experience them and ultimately buy them for their homes as well.

We met the Christofle team for the Asia Pacific, and Olivier Arzel, Managing Director, expert in luxury retail and former director at Cartier. It was very interesting to hear about the brand's expertise, innovation, and design.

"Christofle are experts in quality and innovation with silver. Their range of products is broad but very consistent. It is all about craftsmanship. The forks and knives present the most beautiful details showing the Maison famous chasing and polishing knowledge" says Olivier.

For Christofle, it is crucial to preserve and pass on its silversmithing knowledge. In this spirit, the brand created the Haute-Orfèvrerie range, where focus is given to traditional techniques and handmade production, all done in their workshop in Normandy, France, that currently employs three master silversmiths, who have received the title of "Best Craftsman of France". This type of knowledge makes up Christofle's unique heritage.

Present worldwide in more than 70 countries, after Shanghai (Fig. 4.5) Christofle recently opened in Chengdu, the new Chinese capital of luxury, where people, following the example of the local pandas, know how to enjoy life.

"Christofle style is really between heritage and modernity, and silver is used to reflect all types of decorative arts. The brand has its own design studio in Paris since its inception, and has always been seeking out influential artists, architects, and designers outside. It is with the help of these creative minds that it can remain modern and up-to-date. We can think of Ora Ito designing for the house: his furniture or candelabras are part of Christofle signature pieces. Creations by Gio Ponti or Lino

Sabattini have shaped the history of the brands' designs. Andrée Putman created one of the best-selling collections, which competes with Marcel Wanders Jardin d'Eden line", highlights Olivier. Recently, Christofle teamed up with LaCie to release "Sphere"—a round and stylish silver external hard-drive.

Among other innovations, Christofle invented the "plaqué argent" covering the cutlery with a fine layer of silver, making the cutlery a bit more affordable. More recently, the brand also innovated by creating a process that protects and prolongs the brilliance of silver, for a better polarization. This protective treatment called "Silver Ever" delays the oxidation of silver—a "must-have" for polarotactic luxury shoppers.

4.4.5 Polarotactic Profiles: Business Applications

As many luxury shoppers are polarotactic and therefore attracted to polarized shiny objects like a magnet, the following business opportunities can be grabbed:

– Designing products generating a linearly or circularly polarized light—think of car paint, hair color, or glossy handbags, especially in black and red colors, and of course of natural or man-made diamonds and other crystals
– Commercializing luxury services around polarized lights, like again skiing, sailing, night-clubbing, or scuba-diving
– Organizing events around polarizing moments of the day, like sunset and sunrise
– Taking magnetic fields into account when arranging luxury interior design experiences—at home or in resorts—by implementing relevant feng shui principles for instance
– Paying attention to the lighting with appropriate rotated LED and black light, or on the contrary with natural unpolarized light only

4.5 Designing a Unique and Recognizable Brand

Successful brands are recognizable; no matter the time, no matter the country. They manage to stay true to themselves in spite of designer or management changes, thanks to clear and shared brand codes. We will see in this section how to identify unique brand codes with the Louboutin and the Y-3 cases.

4.5.1 Visible Luxury: Louboutin and the "Chinese Red"

When revamping a brand, isolating the brand DNA and securing distinct brand codes is key. It will guarantee brand consistency over time, facilitate regular updates or market-induced shifts, and lead to a sustainable success. The brand codes' framework will help us identify and share the brand DNA.

To be powerful, brand codes have to be visible. Think of the Apple or LV logo. In that matter, shoe designer Christian Louboutin is a genius. Seeing his assistant putting nail polish on, he got the idea to paint the sole of his high heels red for the season and following the very positive feedback of his customers, kept it as a trademark. Probably having a bit of a shoe fetish, Christian Louboutin first designed shoes for the Folies Bergères dancers and soon his distinct souliers would be seen on popular feet like the ones of Madonna and J-Lo, who even named a song after the brand. For the anecdote, Christian Louboutin got intrigued by shoes as a child while visiting a museum. The museum displayed a sign explaining that high heels shoes were forbidden. Louboutin had never encountered high heels at that time and this is supposed to have triggered his obsession.

Funnily enough, men were the first to wear high heels. Persian cavalry started the trend in the tenth century as heels helped stir the horses and stay on their back. It became therefore a symbol and dress code of the upper-class. Only during the seventeenth century, women also adopted high-heels. And it is very recently that it became a more feminine attribute.

In addition to being distinct, brand codes should be unique and easy to protect. Christian Louboutin is very active in protecting his signature red sole—technically the brand used the Pantone color No. 18-1663 TP also called "Chinese red". The issue is that several courts, in particular in one instance opposing Louboutin to Yves Saint Laurent, who also used a red sole on its shoes, ruled that a company cannot own a color. Some argue maybe Pantone could own it. In that case, I would say that China should own it ;p

So regarding brand codes, luxury brands have to find the right balance between distinct and protectable.

4.5.2 The Brand Codes Framework: The Sofitel Experience

With the example of Sofitel, let's see how to document brand codes.

4.5.2.1 Capturing the Brand Essence

The brand codes framework is a very effective tool to use: easy to understand and to share within the organization. All departments can share the same vision of the brand and it encourages innovation in product development and communication, as the teams know exactly what fits the brand and what doesn't.

The brand codes are also the best way to keep the brand consistent over time. Even if the teams change, the codes will stay. The brand code framework can help identify and highlight the brand assets and make sure to protect them over time and stay true to the brand DNA.

Sofitel luxury hotels cultivated very distinct brand codes. Inspired by its home country France, the brand decided to provide its selected customers with a "cousu-main" (meaning: "hand-sewn") experience. The idea behind this is to pamper the customers and to pay attention to the tiny details that will make the difference.

Each Sofitel showcases its own design, a fusion of French style and local "flair". No matter in which Sofitel though, whether in Amsterdam or Bora-Bora, the crew will welcome you with a friendly "Bonjour!". The Sofitel Munich Bayerpost,

Fig. 4.6 Sofitel Munich Bayerpost (printed with Sofitel permission)

managed by Robert-Jan Woltering—who at the time of printing has been promoted to a new exciting role leading Raffles Hotel in Singapore—is a great illustration of this unique brand experience. The building itself is a beautiful reflection of its surroundings: hosted inside the former Royal Bavarian Post Office and very close to the Theresienwiese, welcoming each year the iconic "Oktoberfest", Sofitel Munich Bayerpost rightly introduces itself as a unique blend of Bavarian traditions, French refinement, and sharp innovation (Fig. 4.6).

Home to many celebrities, the hotel offers 339 guest rooms and as many as 57 suites but, most importantly, an experience. The brand codes of the hotel reflect this attention to details. For instance, the candle ritual brings the typical cozy Bavarian atmosphere, while the MyBed experience, by letting you choose your ideal bedding, finishes to make you feel at home. The Hermès amenities and Haute-Gourmandise sweets served in the restaurant, as well as the ambassadors—each team member represents and conveys the Sofitel experience—will add the French "cousu-main" touch. Life is Magnifique!

With its candle ritual, the Sofitel Munich is very non-polarotactic friendly—an oasis of fuzzy light in a polarized world.

4.5.2.2 Applying the Brand Codes Framework

To build the brand codes, the easiest is to visit the luxury venue, shop, or website. In our case, the Sofitel Munich Bayerpost. Based on our observations, we were able to highlight its features using following multi-sensory descriptors (Fig. 4.7):

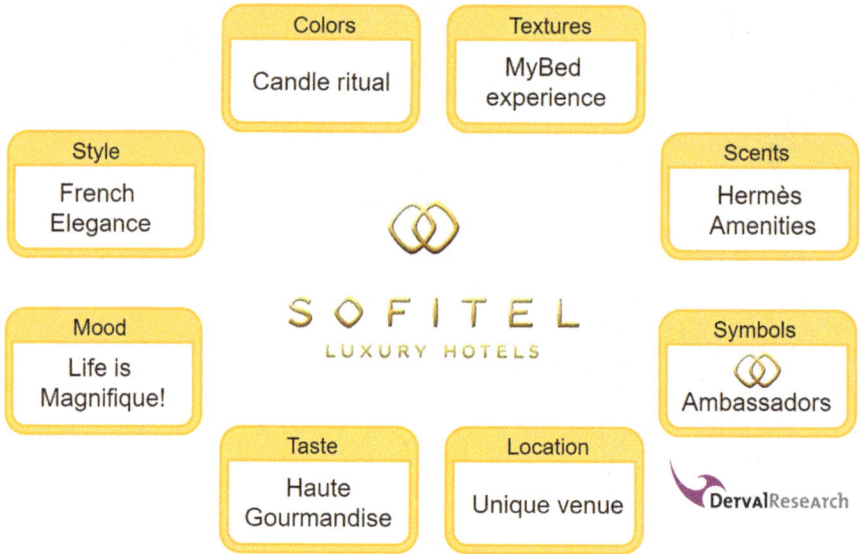

Fig. 4.7 Sofitel Munich Brand Codes (printed with DervalResearch and Sofitel permission)

– Colors: What colors are mostly used? What type of light? At Sofitel Munich Bayerpost, the candle light is iconic and accompanied by a candle ceremony
– Textures: What textures or fabrics are used? We could talk about the fabric of the sofas in the lounge but as it is a hotel and a place where people sleep, MyBed experience is worth mentioning as it allows the guests to select a tailored bedding and pillow
– Scents: What scents or perfumes are perceptible? The Hermès amenities "Eau d'Orange" with a subtle orange scent is popular among the guests
– Symbols: What symbols or icons represent the brand? At Sofitel, the team members are the brand ambassadors. The logo is also a strong trademark
– Location: Does the brand refer to a location? Sofitel adapts to each city and the Bayerpost is so Munich! Still, with a French twist
– Taste: What taste does the brand convey? The specialty is the Haute-Gourmandise—a French version of the high tea, so definitively sweet
– Mood: What is the mood of the brand? With a motto "Life is Magnifique!", enthusiasm is the keyword at Sofitel
– Style: Does the brand show a particular style? French elegance, of course
– Shape: What shapes are being used? A slight preference for rounded corners, like the logo
– Moment in time: Does the brand refer to a particular moment in time? Sofitel would be current and always tasteful

The features can be adapted to the products and services. For instance, in fashion we could add shape, maybe more relevant than taste—even though some showrooms can treat you with a tea or coffee.

So do not hesitate to adapt the framework to your business and project.

Many luxury brands also invest in training the shop floor teams as they are often the first contact point between customers and the brand, and they need therefore to convey the brand codes, once defined.

We had for instance the chance to train Sofitel teams on how to exceed luxury clients' expectations thanks to sensory profiling. At the same time, Richemont group created its own Retail Academy in Shanghai to provide their vendors working for prestigious brands like Montblanc, Lancel, Chloe, Shanghai Tang, or Cartier, with the expertise expected by their target customers. The Richemont Retail Academy is training thousands of sales associates joining the different brands of the group. The objective of the Retail Academy is to select and train the sales employees, and the happy few follow an intensive 8-week program including sessions about etiquette, service, as well as technical aspects about watches, jewelry, fashion, and accessories. DS Automobiles, that we will study in next chapter, also started its luxury training academy.

4.5.3 A Sparkle for Every Smile: The Swarovski Case

When the product portfolio is evolving as in the case of Swarovski, brand codes are critical to stay consistent and credible in the eyes of the customers. Joan Ng, Senior Vice President APAC Product Marketing Jewelry, Watches and Accessories, welcomed us to the world of Swarovski and explained that they can summarize their brand codes in one word: sparkle.

4.5.3.1 Adapting to Local Perception

Leading luxury and premium brands like Chanel and Swarovski seem to propose the same products and services worldwide but when we have a closer look, we will see that they adapt to appeal to local customers' tastes as a Michelin Star Chef would do. Our diamond case revealed the importance of identifying the sensory profile of customers in each target geographical area. Swarovski adapts the products locally. They just released a special collection for India and an Asian collection targeting Japan, Korea, and China (Fig. 4.8). The teams pay great attention to the trends coming right now from Korea. Fine jewelry is in demand in China and the brand plans to develop the Chinese market even further (Derval, 2016).

The innovative Austrian brand, has become the most sought after sparkle brand for consumers and celebrities around the world, including singer-song writer Pharrel Williams. It all started with small crystal figurines and developed into fashion, jewelry, and even wearables. Joan Ng has been contributing to this shift and making sure that the brand adapts to local perception. She agreed to share some insights.

Joan Ng has been with the company for 20 years and took a leading part in shifting it from a crystal figurines to a jewelry fashion brand. Humble, she is proud of her company: "Unlike most luxury brands, who target an elite, Swarovski's philosophy is to propose the best products at an accessible price. Swarovski does not want to exclude anyone. We want to give sparkle to any woman who wants a wonderful shine".

Fig. 4.8 Swarovski jewelry
for Asia (printed with
Swarovski permission)

4.5.3.2 The Extra Sparkle

"What customers love in Swarovski is without a doubt the sparkle. Sparkle comes from the precision-cut crystal and the polish. Our skilled workers polish each jewelry piece by hand to give each angle a maximum of shine. Our brand always paid attention to details and takes pride in having everything checked and checked again. We finish the products with Rhodium—the very expensive and sparkly component that protects against abrasion, which explains the relative higher prices for Swarovski pieces, between 1000 and 6000 rmb (150–920 USD)", shares Joan. Swarovski customers are very Chinese and like the clarity and cut of the stones.

Swarovski has always been very innovative. First, the brand invented a crystal precision cutting technique and more recently introduced its "Activity Crystal" at the Consumer Electronic Show in Las Vegas. Swarovski tracing jewelry collection was the center of all the attention: "Based on the observation that women are both health and fashion conscious, we designed a sparkly fashion accessory collection that can help monitor their activity and sleep, in style."

The brand has its own stores and also partners, retailers who are allowed to sell its products provided that the stores comply with the brand's strict rules. The idea is that in terms of architecture, visual merchandising, and brand experience, customers should not be able to tell the difference. And we are talking about 2300 shops and 8000 multi-brand stores worldwide.

"If a family business is still successful in its 5th generation, it must be doing something right. I think at Swarovski we are passionate at connecting with our consumers and continuously adapting to their demands."

4.6   Take-Aways

Mate Selection

- Mating is all about sending the right signals, and luxury ornaments are considered as signs of health not just of wealth
- Hair, skin, and eyes fluoresce under near infrared light and send special signals, like diamonds, gold, and sports cars

Magnetic Sense

- Luxury items like a red Ferrari, Louboutin shoes, or diamonds are attracting polarotactic people like magnets
- Some people are sensitive to magnetic fields and taking feng shui interior design practices into account can help make a difference

Brand Codes

- Long lasting luxury brands have clear brand codes and they stay true to themselves over time
- Brand codes have to be distinct, and should be protectable
- Documenting brand codes helps reveal and perpetuate the brand DNA
- All successful luxury brands adapt their offering locally even if in appearance they display the same global image

References

Allison, T., Puce, A., & McCarthy, G. (2000). Social perception from visual cues: role of the STS region. *Trends in Cognitive Sciences, 4*(7), 267–278.

Bengtsen, P., & Kelly, A. (2016, July 28). Vauxhall and BMW among car firms linked to child labour over glittery mica paint. *The Guardian*. Retrieved October 15, 2016, from https://www.theguardian.com/global-development/2016/jul/28/vauxhall-bmw-car-firms-linked-child-labour-mica

Brady, P., & Cummings, M. (2010). Differential response to circularly polarized light by the jewel scarab beetle Chrysina gloriosa. *The American Naturalist, 175*(5), 614–620.

Bremer, H. (2017, January 23). *Interview by Diana Derval*. Amsterdam: DervalResearch.

Brown, W. M., Cronk, L., Grochow, K., Jacobson, A., Liu, C. K., Popović, Z., & Trivers, R. (2005). Dance reveals symmetry especially in young men. *Nature, 438*(7071), 1148–1150.

Chiou, T. H., Place, A. R., Caldwell, R. L., Marshall, N. J., & Cronin, T. W. (2012). A novel function for a carotenoid: Astaxanthin used as a polarizer for visual signalling in a mantis shrimp. *Journal of Experimental Biology, 215*(4), 584–589.

Chitrakorn, K. (2016, March 16). Distress in the $80 billion diamond market. *BoF*. Retrieved October 15, 2016, from https://www.businessoffashion.com/articles/intelligence/distress-diamond-price-rout-jewellery-market

Choi, S., Choi, J., Landig, R., Kucsko, G., Zhou, H., Isoya, J., et al. (2017). Observation of discrete time-crystalline order in a disordered dipolar many-body system. *Nature, 543*(7644), 221–225.

Chuang, J. Z., Zhao, Y., & Sung, C. H. (2007). SARA-regulated vesicular targeting underlies formation of the light-sensing organelle in mammalian rods. *Cell, 130*(3), 535–547.

Clark, D. L., & Morjan, C. L. (2001). Attracting female attention: The evolution of dimorphic courtship displays in the jumping spider *Maevia inclemens* (Araneae: Salticidae). *Proceedings of the Royal Society of London B: Biological Sciences, 268*(1484), 2461–2465.

Collins, L. (2011, March 21). Sole Mate. The New Yorker. Retrieved from http://www.newyorker.com/magazine/2011/03/28/sole-mate

Conover, E. (2017, January 23). Chemists strike gold, solve mystery about precious metal's properties. *ScienceNews*. Retrieved from https://www.sciencenews.org/article/chemists-strike-gold-solve-mystery-about-precious-metals-properties

Cronin, T. W., & Marshall, J. (2004). *The unique visual world of mantis shrimps. Complex worlds from simpler nervous systems* (pp. 239–268). Cambridge, MA: MIT Press.

Cronin, T. W., & Marshall, J. (2011). Patterns and properties of polarized light in air and water. *Philosophical Transactions of the Royal Society of London B: Biological Sciences, 366*(1565), 619–626.

Cutting, J. E., & Kozlowski, L. T. (1977). Recognizing friends by their walk: Gait perception without familiarity cues. *Bulletin of the Psychonomic Society, 9*(5), 353–356.

Derval, D. (2010). *The right sensory mix: Targeting consumer product development scientifically.* New York: Springer.

Derval, D. (2016). *Luxury brand marketing 奢侈品品牌营销:创建·实施·案例.* Shanghai: Donghua University Publishing.

Derval, D., & Bremer, J. (2012). *Hormones, talent, and career: Unlock your Hormonal Quotient®.* New York: Springer.

Dugatkin, L. A. (1998). A comment on Lafleur et al.'s re-evaluation of mate-choice copying in guppies. *Animal Behaviour, 56*(2), 513–514.

Fink, B., Weege, B., Neave, N., Ried, B., & Do Lago, O. C. (2014). Female perceptions of male body movements. In *Evolutionary perspectives on human sexual psychology and behavior* (pp. 297–322). New York: Springer.

Fisher, K. J., Recupero, D. L., Schrey, A. W., & Draud, M. J. (2015). Molecular evidence of long wavelength spectral sensitivity in the reverse sexually dichromatic convict Cichlid (*Amatitlania nigrofasciata*). *Copeia, 103*(3), 546–551.

Foley, L. E., Gegear, R. J., & Reppert, S. M. (2011). Human cryptochrome exhibits light-dependent magnetosensitivity. *Nature Communications, 2*, 356.

Gegear, R. J., Foley, L. E., Casselman, A., & Reppert, S. M. (2010). Animal cryptochromes mediate magnetoreception by an unconventional photochemical mechanism. *Nature, 463*(7282), 804–807.

Grammer, K., Fink, B., Juette, A., Ronzal, G., & Thornhill, R. (2001). Female faces and bodies: N-dimensional feature space and attractiveness. *Advances in Visual Cognition, 1*, 91–125.

Greenberg, J., Jordan, R. C., & Sorensen, A. E. (2016). The effect of territory quality on female preference in *Metriaclima zebra*. *African Journal of Ecology, 54*(2), 162–166.

Hahn, C. M., & Campbell, L. J. (2016, August). Birds of a feather laugh together: An investigation of humour style similarity in married couples. *NCBI*. Retrieved October 15, 2016, from http://www.ncbi.nlm.nih.gov/pmc/articles/PMC4991048/

Hanna, J. L. (1987). *To dance is human: A theory of nonverbal communication.* Chicago, IL: University of Chicago Press.

Horváth, G. (2014). *Polarized light and polarization vision in animal sciences.* Berlin, Heidelberg: Springer.

Horváth, G., Gál, J., & Wehner, R. (1997). Why are water-seeking insects not attracted by mirages? The polarization pattern of mirages. *Naturwissenschaften, 84*(7), 300–303.

Houde, A. E. (1988). The effects of female choice and male–male competition on the mating success of male guppies. *Animal Behaviour, 36*(3), 888–896.

Hugill, N., Fink, B., Neave, N., & Seydel, H. (2009). Men's physical strength is associated with women's perceptions of their dancing ability. *Personality and Individual Differences, 47*(5), 527–530.

Hugill, N., Fink, B., Neave, N., Besson, A., & Bunse, L. (2011). Women's perception of men's sensation seeking propensity from their dance movements. *Personality and Individual Differences, 51*(4), 483–487.

Jirotkul, M. (1999). Operational sex ratio influences female preference and male–male competition in guppies. *Animal Behaviour, 58*(2), 287–294.

Johnston, V. S., Hagel, R., Franklin, M., Fink, B., & Grammer, K. (2001). Male facial attractiveness: Evidence for hormone-mediated adaptive design. *Evolution and Human Behavior, 22*(4), 251–267.

Konnen, G. P. (1985). *Polarized light in nature*. CUP Archive.

Kriska, G., Csabai, Z., Boda, P., Malik, P., & Horváth, G. (2006, July 7). Why do red and dark-coloured cars lure aquatic insects? The attraction of water insects to car paintwork explained by reflection–polarization signals. *NCBI*. Retrieved October 15, 2016, from http://www.ncbi.nlm. nih.gov/pmc/articles/PMC1634927/

Lechocinski, N., & Breugnot, S. (2011). Fiber orientation measurement using polarization imaging. *Journal of Cosmetic Science, 62*(2), 85.

Lerner, A., & Browman, H. I. (2016, October 20). The copepod Calanus spp. (Calanidae) is repelled by polarized light. *Scientific Reports, 6*, Article number:35891. doi:https://doi.org/10.1038/ srep35891. Retrieved from http://www.nature.com/articles/srep35891

Manning, J. T., & Taylor, R. P. (2001). Second to fourth digit ratio and male ability in sport: Implications for sexual selection in humans. *Evolution and Human Behavior, 22*(1), 61–69.

Marieb, E. N., & Keller, S. M. (2015). *Essentials of human anatomy & physiology* (12th ed.). San Francisco, CA: Pearson Education.

Mark. (2014, February 20). Fruit, females, and the color orange – Guppy mate preference in the wild. *Corner of the Cabinet*. Retrieved October 15, 2016, from https://cornerofthecabinet.com/ 2014/02/20/fruit-orange-and-guppy-mate-preference/

Marshall, J., & Cronin, T. W. (2011). Polarisation vision. *Current Biology, 21*(3), R101–R105.

Nayak, C. (2017). Condensed-matter physics. Marching to a different quantum beat. *Nature, 543*(7644), 185.

Nießner, C., Denzau, S., Malkemper, E. P., Gross, J. C., Burda, H., Winklhofer, M., & Peichl, L. (2016). Cryptochrome 1 in retinal cone photoreceptors suggests a novel functional role in mammals. *Scientific Reports, 6*.

Patricelli, G. L., Uy, J. A. C., & Borgia, G. (2003). Multiple male traits interact: Attractive bower decorations facilitate attractive behavioural displays in satin bowerbirds. *Proceedings of the Royal Society of London B: Biological Sciences, 270*(1531), 2389–2395.

Place, S. S., Todd, P. M., Penke, L., & Asendorpf, J. B. (2010). Humans show mate copying after observing real mate choices. *Evolution and Human Behavior, 31*(5), 320–325.

Pomozi, I., Horváth, G., & Wehner, R. (2001). How the clear-sky angle of polarization pattern continues underneath clouds: Full-sky measurements and implications for animal orientation. *Journal of Experimental Biology, 204*(17), 2933–2942.

Rapp, L. M., Maple, S. S., & Choi, J. H. (2000). Lutein and zeaxanthin concentrations in rod outer segment membranes from perifoveal and peripheral human retina. *Investigative Ophthalmology & Visual Science, 41*(5), 1200–1209.

Roether, C. L., Omlor, L., Christensen, A., & Giese, M. A. (2009). Critical features for the perception of emotion from gait. *Journal of Vision, 9*(6), 15–15.

Roulin, A., & Bize, P. (2007). Sexual selection in genetic colour-polymorphic species: A review of experimental studies and perspectives. *Journal of Ethology, 25*(2), 99–105.

Rowland, W. J. (1999). Studying visual cues in fish behavior: A review of ethological techniques. *Environmental Biology of Fishes, 56*(3), 285–305.

Roy, G. (2006, June 24). Researchers devise new tool to measure polarization of light. *MyPhys*. Retrieved October 15, 2016, from http://m.phys.org/news/2016-06-tool-polarization.html

Rushton, K. (2015, May 6). A third of the people on hookup app Tinder are already married – With women being most likely to cheat. *Daily Mail*. Retrieved October 15, 2016, from http://www.dailymail.co.uk/sciencetech/article-3070746/A-people-hookup-app-Tinder-married-WOMEN-likely-cheat.html

Singh, D. (2002). Female mate value at a glance: Relationship of waist-to-hip ratio to health, fecundity and attractiveness. *Neuroendocrinology Letters, 23*(Suppl 4), 81–91.

Sommerburg, O., Keunen, J. E., Bird, A. C., & van Kuijk, F. J. (1998). Fruits and vegetables that are sources for lutein and zeaxanthin: The macular pigment in human eyes. *British Journal of Ophthalmology, 82*(8), 907–910.

Staff. (2014). Humans can see infrared light, scientists say. *Sci-News*. Retrieved from http://www.sci-news.com/biology/science-humans-can-see-infrared-light-02313.html

Swirnow, W. (2014, March 1) Properties of emissive materials. *IRInfo.org*. Retrieved from http://www.irinfo.org/03-01-2014-swirnow/

Temple, S. E., McGregor, J. E., Miles, C., Graham, L., Miller, J., Buck, J., & Roberts, N. W. (2015, July). Perceiving polarization with the naked eye: Characterization of human polarization sensitivity. *Proceedings of the Royal Society of London B, 282*(1811), 20150338. The Royal Society.

Troje, N. F., Westhoff, C., & Lavrov, M. (2005). Person identification from biological motion: Effects of structural and kinematic cues. *Perception & Psychophysics, 67*(4), 667–675.

UCSB. (2015). *How does gold get its color?* University of California – Santa Barbara, ScienceLine. Retrieved from http://scienceline.ucsb.edu/getkey.php?key=4367

University of Oxford, Nuffield Health Department of Clinical Neurosciences. (2016). *What is fMRI.* Retrieved from https://www.ndcn.ox.ac.uk/divisions/fmrib/what-is-fmri

Walk, R. D., & Homan, C. P. (1984). Emotion and dance in dynamic light displays. *Bulletin of the Psychonomic Society, 22*(5), 437–440.

Waynforth, D. (2007). Mate choice copying in humans. *Human Nature, 18*(3), 264–271.

Weege, B., Barges, L., Pham, M. N., Shackelford, T. K., & Fink, B. (2015). Women's attractiveness perception of men's dance movements in relation to self-reported and perceived personality. *Evolutionary Psychological Science, 1*(1), 23–27.

Wehner, R. (2001). Polarization vision – A uniform sensory capacity? *Journal of Experimental Biology, 204*(14), 2589–2596.

Wohlsen, M. (2013, May 3). Why Silicon Valley is pouring millions into hair. *Wired*. Retrieved October 15, 2016, from https://www.wired.com/2013/05/why-silicon-valley-is-into-hair/

Zuckerman, M. (2007). Sensation seeking and risky driving, sports, and vocations. In M. Zuckerman (Ed.), *Sensation seeking and risky behavior* (pp. 73–106). Washington: American Psychological Association.

Zuniga, A., Stevenson, R. J., Mahmut, M. K., & Stephen, I. D. (2017). Diet quality and the attractiveness of male body odor. *Evolution and Human Behavior, 38*(1), 136–143.

Expanding Luxury Brands Internationally

5

*"The UAE Space Agency and the Emirates Institution
for Advanced Science and Technology will collaborate to
build the Mars probe."*

HH Sheikh Mohammed Bin Rashid Al Maktoum
(Mohammed bin Rashid Al Maktoum, 2014)

For some reason, foreign brands are always more exotic and desirable. We will see in this chapter, with the Richemont, the Roberto Cavalli, and the DS cases, why expanding luxury brands internationally from the get-go is a best practice. The Wait Marketing approach will reveal a precious help when trying to catch and connect with busy and on-the-go luxury shoppers.

5.1   Introduction

In this chapter, we investigate how to best reach luxury shoppers and via Sect. 5.2 cosmetics case, we raise following questions:

– What is the right time and place to reach luxury shoppers?
– What media and techniques can be used to increase their receptivity?
– How to make the most of luxury tourism and travel retail?
– What is the impact of gifting on luxury shopping?
– How to create positive and long-lasting memories?
– How to design a timeless luxury brand?

In Sect. 5.3, we find out that individuals have a genetical predisposition to traveling or expanding their territory, and that it is linked to their gender polymorphism.

© Springer International Publishing AG, part of Springer Nature 2018   105
D. Derval, *Designing Luxury Brands*, Management for Professionals,
https://doi.org/10.1007/978-3-319-71557-5_5

In Sect. 5.4, we discover that the luxury shopping experience is influenced by biological rhythms and the sense of time, with the Cavalli and DS cases.

We learn, in Sect. 5.5, how to use Wait Marketing in order to communicate with luxury customers at the right moment and at the right place, and review the success stories of Shanghai Tang and Jaguar.

5.2 How to Reach Luxury Shoppers? The Cosmetics Case

We were sitting together in the brainstorming room. A bunch of brilliant managers from a leading cosmetics firm and some of our researchers. The plan was to find the best way to reach out to new target luxury shoppers.

5.2.1 Travel Retail

It all started off quite well with the idea to target travelers at the airport. Surfing on the Wait Marketing concept that demonstrated that consumers are more receptive to interaction while they are waiting as opposed to while they are busy doing a transaction, in which case it is counterproductive to interrupt them, even if very tempting.

London Heathrow airport alone welcomes over 70 million passengers every year, spending on average 1 hour hanging around and taking this opportunity, or not, to do some shopping. Nandita Mahtani, a Mumbai fashion designer, caught at Terminal 5 with a Tiffany shopping bag in her hands justifies: "I've bought some gifts for friends back home. I knew I would buy something at the airport, because I didn't have time when I was in London!" (Neville, 2013). If needed, that proves once again that there is no such a thing as an "impulsive purchase" and "creating needs": some shoppers do not really know when and what exactly they will buy, but one thing is sure, they will buy. Especially in luxury and fashion, where a monthly budget seems already dedicated.

5.2.2 Luxury and Gifting

Traveling is a great time to do some shopping or gifting and to discover and share the brand. Given the layout of the airport—Paris Charles de Gaulle—we needed to be more specific on which consumer groups we would want to target, from which country, in order to open the store not only at the right moment but also at the right place. We started listing the nationalities particularly generous when it comes down to gifting. Many groups popped up to our mind like Americans, Chinese, Russians, Middle-Easterners, and Africans.

In the 60 billion USD travel retail business, beauty products and especially skincare and women's fragrances are in the top sales. The leader is L'Oréal, who even created a dedicated travel retail division, in order not to miss out on this big opportunity. Commuters buy for themselves but also for family and friends. The

French beauty group arranged a special "the perfect gift" campaign around its Cacharel parfum brands Amor Amor, Anaïs, Anaïs, and Noa, with a well thought visual merchandising translated into different languages in order to catch the global gifters on-the-go (Gohel, 2014).

5.2.3   Absurd Decisions

Back to our war-room. We were busy organizing these ideas when one manager abruptly stated "Stop everything! I found the data". Everybody stopped and looked at him. He looked very satisfied and excited. "What do you mean, you found the data?" I wondered. "We are looking for the sales in Paris Charles de Gaulle right? I found the top duty-free sales by nationality! So that's it, we are done", he confirmed with a big smile. The group discussed a bit and approved that we were done indeed and just had to look at this official document published by the airport itself and pick the top three nationalities and basta.

Of a skeptical nature—especially with ready-made reports—I went through the nationalities and something felt wrong, terribly wrong. Some of the ones we listed, like Middle-Easterners for instance, were nowhere to be found in this official list. The group was about to make the most absurd decision in the history of market expansion. What was wrong?

Before we point at the major logical flaw in the reasoning and at the dangers of blindly trusting "big data" or other "fast-thinking" sources of information like Wikipedia, let's explore the relation between people and their territory, why we travel or migrate, and the implications on the luxury and cosmetics industry.

5.3   Gender Polymorphism, Territory, and Luxury

In this section, we will explore the heavy role of hormones on personality, territory, and ultimately on the attitude towards luxury.

5.3.1   Sub-genders and Hormonal Quotient®

Non-human animals are much clearer in their ranking ornaments: male mandrills display a rank-dependent red coloration on their face, rump, and genitals (Setchell & Jean Wickings, 2005), while tree lizards have a blue spot on their throat proportional to their fighting ability (Thompson & Moore, 1991), and red-collared widowbirds a carotenoid status signaling badge more or less yellow or red depending on their morph (Pryke, Andersson, Lawes, & Piper, 2002). I thought for a time that in human animals the tie was playing this role but noticed that even presidents were swapping colors confusingly. Still, you might meet more finance guys into power wearing yellow or red ties and engineers into performance wearing blue shirts than the contrary.

You probably heard of the role of genes and X and Y chromosomes in gender and heredity, but probably less about the role of hormones. Hormones have a triple influence on us with:

- *Intra-uterine programming*, where the pairs of genes we receive from our parents can literally be overwritten by the hormones present in the womb
- *Hormonal organizational effects*, where our physical traits, brains, and sensory perception are set from birth
- *Hormonal activational effects*, where the future effects circulating hormones will have during our lifetime are scheduled in advance (Derval & Bremer, 2012).

Galen, physician to emperors and gladiators, and father of physiology, already noticed that people tended to have different *temperaments*—being phlegmatic, sanguine, melancholic, or choleric. He thought imbalances in *humors* was causing these personality traits. These mysterious *humors* were revealed much later to be hormones and gave birth to a new field of research called *endocrinology* (Brooks, 1962).

A major discovery in that field is *gender polymorphism*. In many animal species, from fishes to lizards, you have different subtypes of males and females. These gender polymorphisms present distinct personality traits, levels of carotenoids, and drivers. In a Texan tree lizard species, for instance, there are three types of males. An orange-throated lizard, who has a very big territory, and is dating several females at the same time. A blue-throated lizard, who has a smaller territory and is dating only one female at a time. And a yellow-throated lizard, who does not stick to any territory, but goes from opportunity to opportunity, and while the two other lizards are fighting, he is flirting with the girls. Talking about the females, two types were identified. An orange-throated female lizard, who doesn't get along with other females and is laying eggs regularly. A yellow-throated female, who gets on well with males and females, and is laying eggs less often but they are bigger. Zoologists found out that the only difference between the lizards, that explained their distinct personality, physiology, and need for status, was the level of exposure to prenatal hormones while in the egg (Derval & Bremer, 2012).

Each and every part of our body reacts to the exposure to prenatal hormones by growing or not growing. In other words, we could know from birth who will have big boobs or be bald, and start saving money for cosmetic surgery (no, but don't do that). Each and every part of our body can therefore act as a biomarker, like a fossil testifying of the levels of prenatal hormones, which is good as otherwise we could only measure the circulating hormones levels and not the more decisive prenatal hormones shaping our body and brain through its "organizational effects". Researchers formally validated that for instance on our right hand, the ring finger has more androgen or testosterone receptors (AR) and estrogen receptors-α (ER) than the index. Inactivation of testosterone decreases the growth of the ring finger, leading to a higher index to ring finger ratio, whereas inactivation of estrogen increases the growth of the ring finger, leading to a lower index to ring finger ratio (Zheng & Cohn, 2011).

We patented a tool measuring this ratio, and based on measurements conducted on 3500 people in 25 countries, we were able to identify 8 Hormonal Quotient® (HQ) profiles (Derval & Bremer, 2012). You can take the test on www.derval-research.com and check your very own Hormonal Quotient® (HQ) profile (Fig. 5.1).

A woman can have been very influenced by prenatal testosterone like Angelina Jolie, influenced by testosterone like Emma Watson, equally influenced by testosterone and estrogen like Kylie Minogue, or more influenced by prenatal estrogen like Victoria Beckham. Now you understand why many testosterone-driven women with Angelina Jolie-like square jaws are having surgery to get a Victoria Beckham-like triangular chin, especially in Asia. It is not so much about improving appearance—I think there are many types of beauty—but more about looking like another polymorphism maybe more popular in a given country because it's rare. If, like me, you are too sensitive for surgery, you can always travel to find a place where your physical traits are better valued (Derval, 2010).

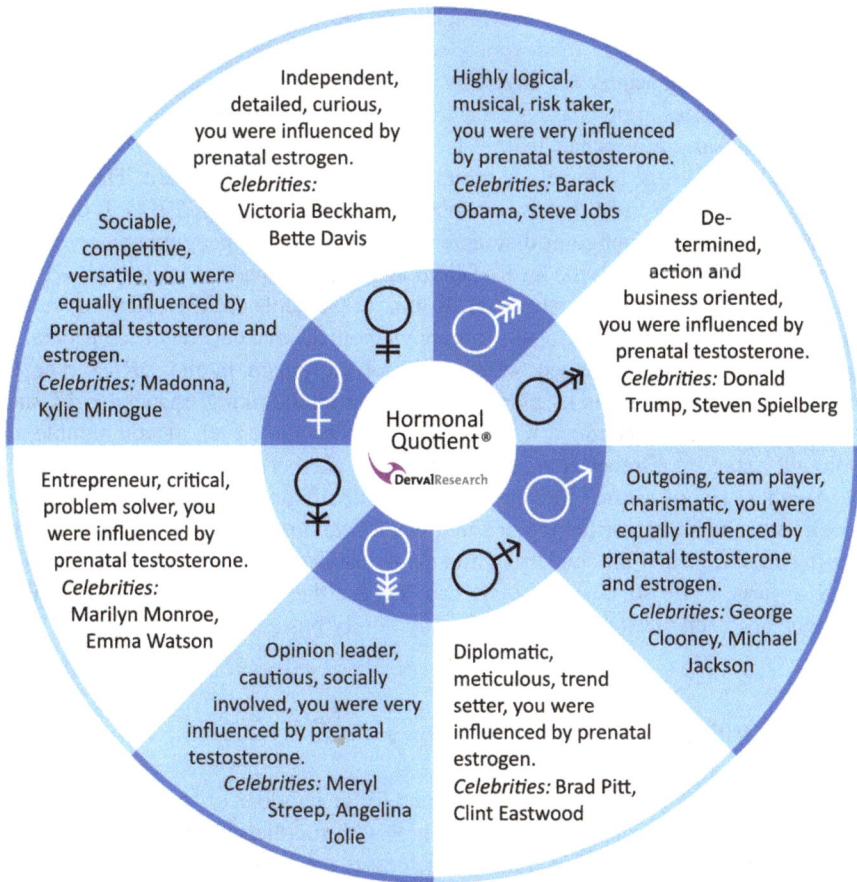

Fig. 5.1 Hormonal Quotient® (HQ) profiles (printed with DervalResearch permission)

If you remember in Chap. 2, the Emirati saw the benefits of the Carl Zeiss expensive lenses because of their more testosterone-driven Hormonal Quotient® generating chromatic aberrations on their retina, while women with an estrogen-driven Hormonal Quotient® would remember their children's drawings as the nicest gift even over the latest designer's bag. Luxury shoppers tend to look up to celebrities with similar or with a more extreme Hormonal Quotient® (HQ), as we will see in next chapter analyzing brand ambassadors and key opinion leaders.

5.3.2 Expanding One's Territory

Expanding one's territory is a strong preoccupation in luxury, and there are many ways leading to this goal, like collecting, traveling, or multiplying.

Some own 50 pairs of shoes (!)—this is way more shoes than they have feet. A series of in-depth interviews conducted on 42 acquisitive buyers, mostly female, owning over 50 pairs of shoes, generated clear insights on the buying motivations—related to materialism, perfectionism, and variety-seeking (Bose, Burns, & Garretson Folse, 2013). Interviewing acquisitive buyers as well as non-acquisitive buyers helped highlight the differences in purchasing behavior between the two groups and how predictable the acquisitive buyers can be once their motivation is established.

Some have an urge to travel, or even to move, like a yellow lizard. The territory we choose to live in truly shapes our social life (Wolch & Dear, 2014). It has been demonstrated in King penguins that increasing the number of penguin neighbors and their proximity could raise the level of circulating cortisol and associated stress by as much as 40% (Viblanc, Gineste, Stier, Robin, & Groscolas, 2014). Rethink before moving to a big city! Here again, our physiological makeup has a considerable power as just being equipped with a 7R polymorphism in the DRD4 type of dopamine receptors makes us more prone to migrating, taking financial risks, and suffering from food disorders. It's a bit like whether you travel, or you gamble, or you will binge eat (Dreber et al., 2010; Derval & Bremer, 2012).

Another "call of nature" for some is to generate offspring. I think my "biological clock" must be broken or ticking in a frequency I cannot hear, but it is fascinating to see my entourage multiply and try to find many justifications to it other than they just have an urge to do so. In luxury, we can observe that filiation and transmitting a heritage has been the starting point of many family businesses.

5.3.3 Affiliation, Family, and Luxury

Luxury firms are often family firms, installed for generations on their territory, with strong roots in rituals, and strong relationships with suppliers and partners (Carcano, Corbetta, & Minichilli, 2011). Family businesses are very active in luxury, retail, automotive, and media, and constitute the third leading business force in the world in terms of added value creation (Peterson-Withorn, 2015).

Very successful family owned businesses include Louis Vuitton, Hermès, Gucci, Chanel, Hennessy, Cartier, and Fendi, to name a few (Carcano et al., 2011). Family businesses are above average at combining key luxury ingredients such as a. continuity, with timeless pieces such as the Hermès Birkin bag for instance, b. community, creating a multi-competency team of designers and artisans working together, c. connections, rooted in their territory, family luxury brands have developed long-lasting and strong relationships with their partners and suppliers, and d. command, as family management type is often based on passion and fast decision-making resulting in bolder and more creative orientations (Miller & Le Breton-Miller, 2005).

We can think of watches transmitted from father to son or jewelry from mother to daughter. Filiation is a way to expand territory over time.

5.4 Luxury Is Timeless

Luxury is not just about shiny products, it encompasses also precious memories. Whether dining at a Michelin star restaurant, or attending the Fashion Week, the way we perceive and memorize experiences is key. The perception of time, the biological rhythms, the memories of places and events, are closely linked to multisensory stimuli.

5.4.1 The Sense of Time

The biological rhythms, time and place receptors, and other time modifiers, impact our behavior and shape our luxury memories.

5.4.1.1 Ultradian, Circadian, and Infradian Rhythms

Life and even fashion follows a cycle: the 90's can for instance become trendy again. This is because non-human and human animals tend to follow cycles. We can think of seasons, lunar cycles (lasting 12 hours and 25 min each), tides, or days. Crabs would typically stay on the beach a time equivalent to two and a half tides.

People tend for instance to eat more, have more sex (with a peak at 11pm so that you know), and sleep more during the week-ends. They are also more likely to suffer higher blood pressure, or a heart attack on Mondays (Refinetti, 2016). While working in the travel business, I was also able to observe thanks to the connections to booking servers that most people arrange their holidays and week-ends when they arrive at work on Monday, as it seems too depressing to start working right away.

In India, October and November (during Diwali) are the golden jewelry months with a strong increase in sales linked to all the weddings planned during the non rainy season (GCI, 2013). In the U.S., Christmas is the clear diamond season. Married women receive, or expect to receive, at that time, their shiny treat, followed by Valentine's Day and Mother's Day. Middle-Easterners tend to go on diamond safaris in Europe shortly before Ramadan (Laniado, 2015).

Luxury shoppers and fashionistas are very sensitive to seasons, as they like to be versatile, and some will buy multiple accessories to fit into any situation. We will see later in this section that temperature, food intake, and many other factors influence our time perception and management, and in the end our luxury and fashion shopping behavior. Human cycles create patterns over a period of time that can be infradian, when the cycle is every 28 hours or more, circadian, when the cycle lasts between 19 and 28 hours, and ultradian, when the cycle is shorter than 19 h (Refinetti, 2016). Circadian cycles are rhythmed by the day, the night, and two twilights: dawn and dusk—moments with high polarized light as we studied in previous chapter. They drive motion, feeding, body temperature, sensory processing, and learning abilities. Ultradian cycles are influenced by growth hormones (Tannenbaum & Ling, 1984), and are linked to leptin and obesity (Tolle et al., 2002), so that ultradian oscillations of insulin make diabetics more impatient. So if you are selling to sugar-addicted customers, you better not have them wait too long!

Annual cycles organize in some species food intake, melatonin secretion, reproduction, and even car accidents. Testosterone and sex hormones also fluctuate in cycles, which explains a peak of car accidents in summer (Reinberg, Lagoguey, Chauffournier, & Cesselin, 1975). Animals with a dysfunctional circadian clock display oscillations in periods of 2 to 5 hours, instead of 24 hours. Also, in absence of light and dark references, people will have an active cycle of 16 hours, like flies and geckos, against 12 hours for an iguana lizard, and 10 hours for a tarantula (Refinetti, 2016).

Movement creates sound as we saw in Chap. 2, and it is no different with our cycles. Infradian cycles oscillate in a frequency of 9 μH, circadian 12 μH and ultradian 16 μH. The circadian clock is also known as our internal pacemaker. A horse heart will tick less than one time per second and a rat heart 5 times per second, like a vintage Rolex (Refinetti, 2016).

5.4.1.2 Time and Place Neurons

Dr. John M. O'Keefe, Dr. May-Britt Moser, and Dr. Edvard I. Moser received, in 2014, the Nobel Prize in physiology for discovering, in the brain, cells that give us a sense of place and navigation (Burgess, 2014), These place cells, located in the hippocampus, help build a *spatial map* of the environment. Put together in grids, in another part of the brain—the entorhinal cortex—they form a navigation tool, our inner GPS. What was discovered shortly after, is that not only place, but also time is encoded in these pyramidal neurons, and they form our memories (Fig. 5.2). It's a pity the pyramidal neurons were not discovered by French scientists as it is obvious they look like the Eiffel Tower—I am happy to be able to highlight it in this book.

Place cells and time cells are the same neurons, so the cells can encode both dimensions together or separately. Research shows that most hippocampal cells contain a time and distance stamp. In addition, they can encode other sensory stimuli like scent or direction (Eichenbaum, 2014). It's very handy to remember the path and navigation. I must be part of the people who remember landmarks or Points of Interest (POIs) and will give you the way with indications like turn left at the supermarket, rather than take the 3rd road on the left.

Location grid

Sensory input

Time and place neuron

DervalResearch

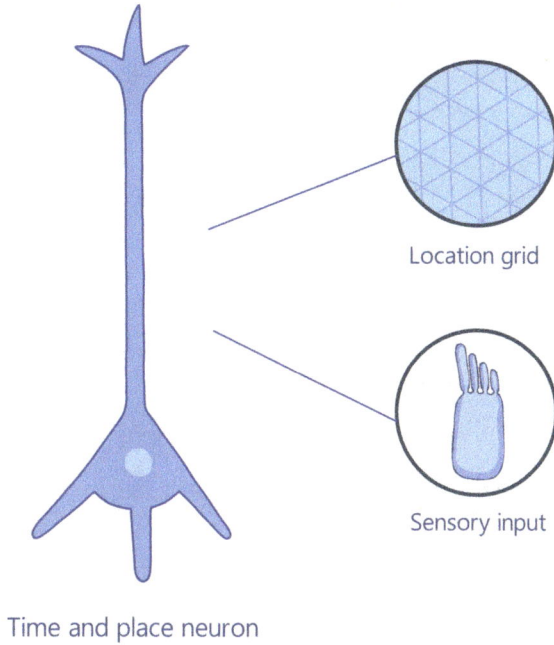

Fig. 5.2 Time and place neuron (printed with DervalResearch permission)

Similarly to place cells being controlled by spatial cues like landmarks and structures like the architecture, time cells are controlled by temporal cues like the beginning and the end of an event, and structures like the interval of an event (Eichenbaum, 2014).

Memories are therefore linked to navigation, but also to the multisensory input received at that moment. Each particular memory is rated as positive or negative and associated with the place where it got memorized.

Time and place are intimately connected as shown by the Kappa effect: the time interval between two stimuli—in the experiment, it was a source of light blinking—seems longer when the stimuli are coming from different locations than when they come from the same place (Quinn & Bausenhart, 2014).

5.4.1.3 Time Modifiers

Our perception of time is influenced by factors like our pulses, light, sound, attention, energy, mood, temperature, pain, familiarity, and memory (Fig. 5.3). An increase in our body temperature or a sensation of pain leads to longer duration estimates. Pulses are like an internal feedback, similarly to the feeling of satiety (you know, the signal confirming you ate enough that less skinny people like me obviously do not receive in a timely fashion, just saying). The perception of time

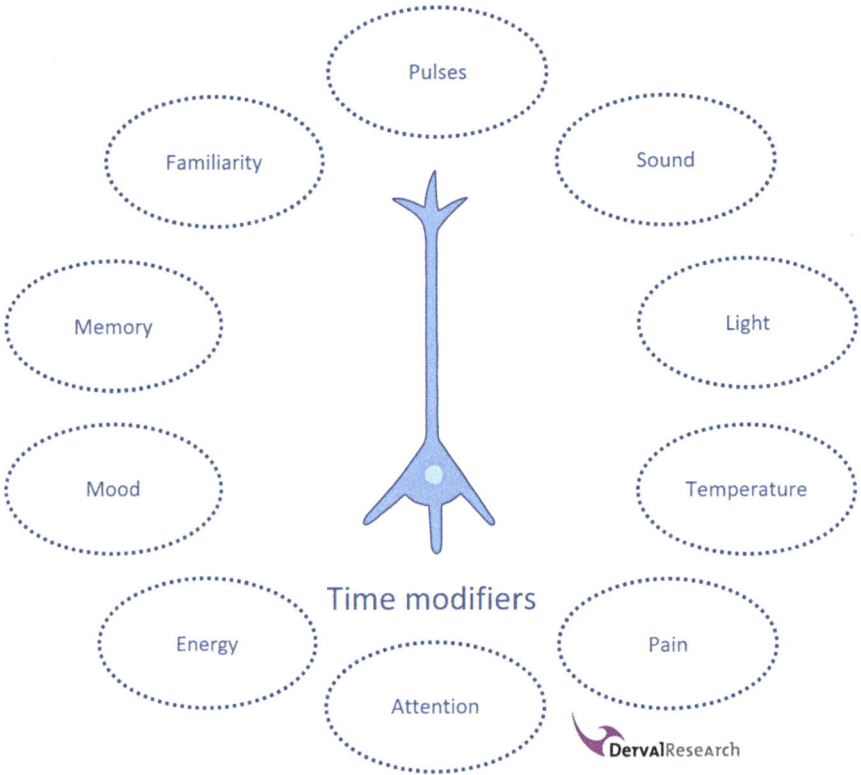

Fig. 5.3 Time modifiers (printed with DervalResearch permission)

is therefore somehow related to proprioception, which we studied in Chap. 2, and how we feel our inner organs. In other words, good sailors probably have a much better sense of time passing by than clumsy nerds like me (double pleonasm).

In spatial perception, the visual system seems to dominate, whereas in time perception (Bertelson & Aschersleben, 1998), the auditory system seems to dominate (Romei, De Haas, Mok, & Driver, 2011), highlighting the importance of sound and music. When we listen to music, we automatically switch to an auditory prediction mode, where we anticipate the next beat based on the previous ones. Perceiving the internal pulse definitely helps. Now I remember that my sister who used to play drums in a band had to listen to a tune with a beat. Most likely our sailor with strong proprioception has the tune constantly in his head. Based on tests, the median tempo would be 133 beat per minute (BPM), more or less 30% (Patel & Iversen, 2014).

Humans are able to very closely follow the beat of music—and have in that extent similar talents to parrots, Asian elephants, and California sea lions—because in fact they anticipate the next beat in a skilled temporal prediction (Patel & Iversen, 2014).

When asked to tap fingers at their favorite beat, a tempo of 2.3 taps per second was preferred (Refinetti, 2016).

Perception of short-term and very-short term are handled distinctly, so that waiting for a couple of milli-seconds has nothing to do with waiting for a couple of seconds. The frontier is supposed to lie around 200 ms. We emit our own pulses at regular interval and measuring these pulses gives us an evaluation of time passing by. For that reason, and because we do not perceive the other contextual factors in the same way, two people can have a completely different perception of the same period of time (Wittmann, 2013). Familiarity will also distort memories: a familiar territory is perceived as being bigger and a familiar commuting time as being shorter (Jafarpour & Spiers, 2017).

The pulses have been compared to the ticks of a watch. If you recall from Chap. 2 with the Jaeger-LeCoultre case, some individuals would prefer wearing a 36,000 vpm watch while others are happy with a 21,600 vpm. We can bet that people synchronize with their watch and therefore pick the right tick. For someone without any notion of time, having a thing ticking at regular intervals is simply nerve wracking: since you do not properly sense any internal pulses, the watch ticking is not perceived as regular but just as random and constantly annoying.

Mood and emotions alter our perception of time: time slows down with fear for instance (Refinetti, 2016). Like in movies, we switch to slow-motion mode, probably to give us more chances to assess the situation and react appropriately. Boredom, too, of course.

5.4.1.4 Memory Profiles: The Bear, the Monkey, and the Banana

Memory plays a key role in our evaluation of time. The way we store information in our brain has a direct impact on the type of memories we retrieve and this little test will help you find out.

Here's a bear, a monkey, a banana: How would you split them into two groups? (Fig. 5.4)

Fig. 5.4 Bear, Monkey, Banana test (DervalResearch, printed with permission)

Make your decision and then check the results (Derval & Bremer, 2012):

(A) Bear and Monkey | Banana

People grouping without any hesitation the bear with the monkey have their brain organized into categories, like animals, or fruits. They have strong analytical skills and are fast at processing data, as they are all already organized. Some might find them a bit narrow-minded and resistant to change as change might require building new categories, but once the new boxes are in place they are true change champions. The way of thinking is sequential and linear. So memories can only be retrieved one way—the way they have been stored.

(B) Monkey and Banana | Bear

People grouping without any hesitation the monkey with the banana, because one eats the other, store information together with their context and attributes, and all attributes are indexed and linked to each other. So, for instance, the monkey is brown like a bear on an iceberg on which you can slip like on a banana eaten by a monkey. Some might think they are following too much their intuition or instinct as they can have difficulties recalling and explaining their whole thinking path that is logical but not sequential. The way of thinking is holistic or indexed. People can easily compare situations and come to new ideas by analogy and retrieve memories from multiple angles.

(C) Bear and Monkey | Banana OR Monkey and Banana | Bear

People hesitating between grouping the monkey with the bear or with the banana store the information into categories, but at the same time link them with indexes. They can hear and understand different points of view and are therefore facilitating the flow in projects. Some might think it takes them too long to make a decision. The way they think is both sequential and indexed. They can retrieve memories from several angles.

(D) Bear and Banana | Monkey

People grouping the bear with the banana are, or trying to upset the test-maker (but, of course, it didn't work as I anticipated it all!) which denotes a certain tendency to be a rebel, or they are trying, believe it or not, to save the banana! But who asked you to save the banana? In fact, people in that case store information in categories like in (A) but with an extra category called "valuables" where they store food and gold. We did the test on many people and this answer mostly came up among finance and banking teams, so that you know (Table 5.1).

With the help of oxytocin, important information is separated from background noise. When forming memories, oxytocin filters the input coming from pyramidal neurons and interneurons. This is why people a bit lower on oxytocin, for instance those on the autism spectrum (like me, for example!), have a photographic memory, as nothing has been filtered while storing information (Eichenbaum, 2014). Also, being a bit of a nerd makes you less subject to illusions, as multisensory integration works differently and most sensory signals are received separately instead of being

Table 5.1 Memory profiles (printed with DervalResearch permission)

Way of thinking	Sequential	Sequential indexed	Indexed
Brain organization	Has the brain organized into categories, like animals or fruits, and sometimes even with a special category like "valuables"	Stores the information into categories, but at the same time links them with indexes	Stores information together with their context and attributes and all attributes are indexed and linked to each other
Bear, Monkey, Banana test	Bear with monkey without any hesitation or bear with banana	Hesitating between grouping the monkey with the bear or with the banana	Monkey with banana without any hesitation
Strengths and weaknesses	Has strong analytical skills and is fast at processing data, as they are all already organized. Some might find him/her a bit narrow-minded and resistant to change as change might require to build new categories, but once the new boxes are in place he/she is a true change champion	Can hear and understand different points of view and is therefore facilitating the flow in projects. Some might think it takes him/her too long to make a decision	Some might think he/she is following too much his/her intuition or instinct as he/she can have difficulties recalling and explaining the whole thinking path that is logical but not sequential. Can easily compare situations and come to new ideas by analogy
Memories	Memories can only be retrieved one way—the way they have been stored	Memories can be retrieved from several angles	Detailed memories can be retrieved from multiple angles
Luxury brands	Louboutin, LV	Hermès, Shanghai Tang	10 Corso Como, DS

Based on measurements and observations performed by DervalResearch on 1200 luxury and non-luxury shoppers in over 25 countries from April 2012 to March 2017

integrated. In other words, no matter the light, food will taste the same and dresses will still be blue (Russo et al., 2010).

Let's see how family brands like Michelin and DS successfully associate with great memories.

5.4.2   Dining with Michelin Stars: The 10 Corso Como Case

Every year secret tasters visit restaurants and judge the quality of the food and dining environment granting from zero to three coveted Michelin stars to the Chef. Fascinatingly, it's the same Michelin Man who helps you find your way, change your tire, and serves you food!

Michelin is a great family business success story. The brand opted for lifestyle instead of a fixed product category. With the positioning "moving forward together", the French firm commercializes, worldwide, tires, maps, sightseeing guides, navigation tools, and the famous fine-dining reference, the Michelin Guide. In the case of safety-related products like tires and food, the whole brand image has been built on the perception customers have of Michelin's quality. What is sure is that many brands tried to outsmart the Michelin Guide but with little success. A portfolio based on a lifestyle rather than on a product expertise enables Michelin to play on very different markets without confusing customers.

The initial vision of Michelin's founder, from Clermont-Ferrand in France, at that time producing tires, was to give drivers more excuses to drive—and by doing so, to use their tires!—by providing them with scenic routes to visit and dining places to go to. A brilliant implementation of the Wait Marketing technique, turning luxury consumers' available time into business opportunities. Only the elite was able to afford cars at that time, which is why the Michelin Red Guide features mostly high-end restaurants.

To learn more about luxury restaurants and fine dining, we met with Chef Corrado Michelazzo, in charge of 10 Corso Como restaurants in Beijing and Shanghai. 10 Corso Como is a luxury lifestyle department store combining arts, architecture, fashion, and food, with an Italian flair. Located originally in Milano, then Seoul, Beijing, and Shanghai, the brand was founded by Carla Sozzani, former editor-in-chief at Vogue and Elle Italy.

Chef Corrado agreed to tell us how it feels to get a Michelin star: "I did my cooking classes with famous chefs like Troisgrois and Marc Veyrat and quickly got my first Michelin star only 2 years after opening my kitchen in Italy. The star brought a lot of additional responsibilities. So this was an exciting but also very tough time, as I was managing six or seven projects at the same time. I lost too much weight back then!" Now Chef Corrado is Chef at 10 Corso Como restaurant in the Jing'an area in Shanghai, and busy with the opening of a second cuisine in Beijing. He has more time to focus on the menu. The current menu is a fusion between Italian and Japanese cuisine: "A cuisine between earth and sea elements. I am passionate about molecular cuisine and fine ingredients. Earlier this year we launched for instance a five-course menu including quality Chinese products like Yunnan wild boar, Yangcheng Lake crab, and Hangzhou Longjing tea", shared Chef Corrado. The Chef also developed options for businessmen and women working in the area. His secret ingredient to success in such a competitive field as gastronomy resides in creating stunning dishes by using analogy. The Chef gets inspired by daily life activities: "A mojito can suddenly make me think of a new fish meal with a hint of mint." Definitely a "monkey with banana" type of memory here.

While preparing this book we had the chance to taste a special 10-course menu prepared by Chef Corrado Michelazzo and we were enchanted—and trust me, we have quite some taste buds in the team.

Luxury shoppers enjoy more and more having dinner in a Michelin-star restaurant or spending quality time in a super luxury resort. Luxury is all about the lifestyle and the precious memories and not only the products anymore.

5.4.3   Fashion Week: Derval Paris Haute-Couture Hats

The Derval Paris case gives a great example of seasonality and Wait Marketing applica-tion in luxury fashion. When my team and I decided to revamp the Derval Paris brand, we did a lot of research in order to re-collect the symbols, codes, and anecdotes of Derval and bring its true story to life. What we did was close to archeology and what we found was quite exciting. We went back to the Derval Castle in France to get inspired. It was incredible to see how splendid the fortified castle was after over 1000 years history! This popular place of leisure since medieval times, with luxurious gardens, resisted so many assaults, and inspired us with a both "romantic" and "fierce" image.

Derval history starts in France at the end of the tenth century with 'Austroberte' and the Lords of Derval. Their symbol is the stoat and their saying "sans plus" means that no one can top them. Jean de Derval, the most powerful Lord, restored many castles, and received high distinction from the king for his actions.

The fortified castle of Derval, famous for withstanding the assaults of many invaders, quickly became the popular place for Ladies and Lords of Derval to enjoy leisure time and "la vie de château" (the castle lifestyle). See how the Derval Castle looked like in 1375, in the beautiful painting found in Pierre Le Baud's book on Brittany (Fig. 5.5).

With fierce and romantic hats and accessories, the brand particularly appeals to performers, fashionistas, and high achievers. Derval Paris is attracting luxury personas who are more into performance and originality than into power. With the help of a team of designers, selected from top fashion schools including IFA Paris, FIT, Marangoni, and Donghua Shanghai International College of Fashion and Innovation, the accessories are customized to meet the specific needs and style of each customer.

Each piece is handcrafted following ancestral techniques of carving, embroidery, pokerwork, and braiding. I like to compare Derval Paris with Ladurée: "Derval Paris fashion accessories are like Ladurée macarons. Made of simple ingredients, it is the passion and sense of detail that turn them into exquisite creations".

Therefore, as a brand we decided to link our haute-couture hats to another unique field: arts. The first collection was featured at Shanghai Museum of Contemporary Art (MOCA), the following collection was inspired by a Rembrandt painting, and the perfume "La Vie de Château" by the Derval Castle painting.

During Shanghai Fashion Week, using the Wait Marketing approach, Derval Paris revealed a horny hat (see Fig. 5.6), inspired by Peter Paul Rubens' painting of the Two Satyrs executed in 1618—in the mythology, satyrs were lustful woodland gods with goat's ears, tail, legs, and horns. As all the fashionistas and media in town were at the entrance of the Fashion Week venue in Xintiandi area, we had our Dames de Derval (understand Ladies of Derval) make an entrance wearing the new hat collection. Also, celebrity fashion photographer and writer Wanglili, a true hat lover, is always keen on wearing our new creations and became a brand ambassador. In that way new collections are in the spotlights during the whole Fashion Week.

Fig. 5.5 Derval Castle in 1375 (Le Baud, 1484, XV, printed with BnF permission)

To convert the target luxury customers, not only must the brand codes be spot on, from the story to the international flair, but also brands need to approach their customers at the right moment, at the right place.

Fig. 5.6 Derval Paris horny hat (printed with Derval Paris permission, photo 肖光Xiaoguang)

5.4.4 Selling Cars Like Luxury Shoes: The DS Case

That is exactly what new luxury car brand DS Automobiles is doing. DS, partner of the Louvre Museum since 2015, also aspires to combine French tradition and avant-gardism under the lead of its CEO Yves Bonnefont.

Like other premium car brands, DS partnered with watch makers and Formula E teams, with the DS Virgin Racing team. What is far more unusual is the partnership with Givenchy—the French luxury fashion and perfume brand—on the city-friendly DS 3 model or the choice of the colorful and 94 year-old eccentric fashion designer and icon Iris Apfel.

DS is now one of the newest and notably prominent premium car brands in China, and their increased sales are without a doubt due to the smart sponsorship of the Shanghai Fashion Week, the car itself, and to the brand ambassador Sophie Marceau singing the Piaf song "La vie en rose" live on CCTV during New Year's eve.

With two DS World showrooms in Paris and Shanghai, 132 DS Stores, and 95 DS Salons including one in Tehran, the brand truly innovated again with its DS Urban Stores. The brand's legendary blue 1972 DS 21 was revived in the recent The Mentalist CBS TV series, and DS just opened its first DS Urban Store in London's Westfield luxury shopping hub, right next to Gucci and Burberry (Fig. 5.7). The vision is that women might want to touch, compose—with the help of a car configurator—and buy their unique DS, like they would buy a pair of shoes—luxury shoes, of course.

Fig. 5.7 DS Urban Store in London-Westfield (printed with DS permission)

In this unique space of 140 m^2, a team of advisors will help you immerse into the brand with a test drive or a 3D Virtual Tour powered by Dassault Systems, showcase for you some of the "Lifestyle" items, or help you clinch a deal in the more quiet Bijouterie corner ("bijouterie" means jewelry store in French). Making the most of the Wait Marketing approach and connecting with shoppers while they are available browsing their favorite brands, DS is keen on expanding the DS Urban Store success to other locations (Fig. 5.8).

Yves Bonnefont agreed to share some insider information with us (Bonnefont, 2017). DS has been cultivating a "spirit of avant-garde" since its inception in 2015, when it became a brand inspired by the iconic DS model, comparable to the Ford Model-T or the VW Beetle. Combining refined materials, French know-how, and the latest technology, the brand showrooms and boutiques align on luxury hotels to provide DS drivers with a valet service arranging all maintenance-related requirements—picking up and delivering cars to their homes. In an indeed very avant-garde approach, DS considers all DS drivers their precious clients—whether they bought or are renting a DS, it doesn't matter, they will be equally welcomed at the Club Prestige DS. Yves explains: "We studied our personas and tried to find some wow factors to impress them but they were more into simple services helping them to fully enjoy the DS experience like for instance the DS Valet Parking, where a voiturier can be booked 15 min in advance to park your DS in busy cities". The DS Urban Store is part of this luxury service vision—engaging with current and potential customers at the right place and at the right moment—as well as the DS Academy in charge of training all the DS teams worldwide to these luxury service standards—quite unusual in the car industry—across the 200 DS points of sales, to

Fig. 5.8 DS Urban Store in London-Westfield (printed with DS permission)

be multiplied by 4 within 2 years. "Luxury is first and foremost an emotion. DS attempts to create this emotion through the beauty of the product design, the refinement of the materials, and a handmade touch, as all the steering wheels are hand-sewn", highlights Yves Bonnefont.

5.4.5 Importance of Being Global: The Charles Philip Case

How could an individual obsessed by expanding his/her territory possibly trust a brand only implanted locally? The answer is, this cannot work, as we will see with the Charles Philip case in Shanghai.

Charles Philip observed how Chinese women did not feel comfortable wearing high heels and built a luxury slippers emporium for China. He agreed to share his global experience.

Charles Philip, founder of Charles Philip Shanghai, grew up in Milan and lives now in Shanghai. He and his team had the vision that many people would enjoy flat shoes in China. To make it in China, they felt it was key to first become popular outside of China, and managed to get distributed at Saks Fifth Avenue in New York. Each season their slippers designed and made in China are offered in over 50 colors and patterns for around 200 USD per pair—the white and blue stripes being the signature pattern (Philip, 2015).

Charles Philip Shanghai was the very first luxury brand to mention Shanghai in the brand name instead of Paris, Milano, or New York. You could tell that the Chinese were very sensitive to it. The company even started its own production in

Shanghai, with skilled workers and a dedicated design studio, in order to have full control over the quality and ensure customers all products are 100% made under the brand direct supervision. The shoe workshop became a strong selling point and is showcased in social media. Given the size of the domestic luxury fashion market in China, brands do not even need to export for the revenues, but it is a must in order to prove to Chinese customers that even foreigners acknowledge the brand. When the Chinese First Lady Peng Liyuan was wearing an Exception dress and handbag, while representing the country abroad, it instantly increased the demand for the Guangzhou-based brand. Therefore it is key to build a global luxury brand up front, and get this critical international endorsement. Research conducted among 1000 Chinese consumers across 14 cities confirmed that the top 3 factors for buying luxury items have shifted. If superior craftsmanship is valued for fashion, leather goods, jewelry, and watches, buying from an internationally well-known brand is confirmed as being a must (Derval, 2016).

One third of luxury goods are purchased by Chinese consumers, particularly excited by powerful cars and luxury fashion items. In spite of a large market, a local expertise in fashion and textile—20% of Prada products are created in Chinese workshops—and a history in luxury products with silk, or porcelains, luxury consumers, and even the Chinese ones, favor western brands. Moreover, due to high taxes on luxury products, most purchases are made abroad. A "proud to be Chinese" movement made its appearance among the young elite and many Chinese study abroad to learn from western countries and come back to China to contribute to the growth of their homeland. The challenge for China is to replace the image of being the world's workshop by being a fashion capital in the same way as Paris, Milano, and New York. Most subsidies are right now focused on automotive and energy firms but the government and local authorities certainly play an important role in the luxury market, by supporting the Fashion Week for instance.

The other challenge of having more products designed in China is adapting the hierarchical management style to handle designers and to train sales vendors to provide shoppers with the right level of service and product expertise. New luxury brands would benefit from targeting the areas where existing, mostly Western, brands did not properly address consumers' morphology, perception, and needs.

Being successful abroad and creating a lifestyle following the lead of Shanghai Tang, we will study shortly, is a must, as it gives the brand credibility and helps promote the "Chinese Dream" and "designed in China" label. Chinese brands are overpowering their western counterparts in various industries from telecom to social media but the luxury market is definitely a tough cookie. Surfing on the "proud of being Chinese" wave, some shopping malls like Xintiandi Style in Shanghai put Chinese designers under the spotlight. The Taiwanese-owned mall dedicated the whole second and third floor to Chinese designers and their names and hand prints are even featured on a wall of fame. Among them is Helen Lee, the rising star of recent Shanghai Fashion Week.

If some local brands emerge and the mindsets seem to change, it is still a challenge for China to turn from a workshop into a fashion capital. Organizations like SHIFF, Shanghai International Fashion Federation, are working hard on it by creating for instance a Chinese Haute-Couture cluster and helping designers meet the demanding standards in that field. Renowned Donghua University, specializing in fashion and textile—the Chinese spacesuit has been designed there!—also puts efforts into creating SCF Shanghai International College of Fashion and Innovation in partnership with prestigious Edinburgh University to offer a top education in fashion design and fashion marketing and help boost the creativity in Shanghai.

Designer Niki from Moodbox creates the event at each Shanghai Fashion Week with inventive and fashion-forward pieces that TV hostesses, superstar singers MoMo Wu, and powerful business women are falling in love with. With a boutique in Xintiandi Style mall in Shanghai, designer Niki managed to propose avant-garde and at the same time wearable pieces for both women and men.

Some mainland brands like Charles Philip Shanghai and PAN's we will discover in Chap. 6 have been particularly successful by understanding local needs.

5.4.6   Memory Profiles: Business Applications

The urge of luxury shoppers to expand their territory, to transmit something to the next generations, as well as the perception of time inspires many business opportunities:

– A family-oriented positioning like Hermès resonates for luxury customers
– Timeless items like a Rolex watch or a fine piece of jewelry or cutlery you can pass on to your offspring are a must
– Services around traveling and international real estate are very appealing to luxury shoppers who need to expand their territory
– Associating luxury brand image with exciting memories like the Fashion Week
– Associating luxury brands with artistic events and creations in order to leave marks over time.

5.5   Reaching Luxury Shoppers with Wait Marketing

We saw with the Personas framework that it is critical to target the right luxury shoppers, and we will see now, with the Wait Marketing approach, and the Cavalli, Jaguar, and Richemont cases that targeting is great but not enough: you need also to communicate at the right moment, at the right place.

Ancient Greeks were very organized. They had a god for everything. You have probably heard of Kronos, their god of time, but have you heard of Kaïros, their god of the right moment and of the good opportunity? Legend depicts Kaïros as a good-looking guy, with long hair and wings on his feet (Fig. 5.9). Legend says, when he passes by, very fast, you have three options:

Option #1: You do not even see him
Option #2: You see him and do nothing
Option #3: You see him and catch him by the hair, grabbing the opportunity

Fig. 5.9 Kaïros, the god of
Wait Marketing (printed with
DervalResearch permission)

RECOMMENDED BY KAÏROS
GREEK GOD OF THE RIGHT MOMENT

I hope reading this section will make you want to grab the opportunity of Wait
Marketing, and save 30–70% on your communications budget (Derval, 2007).

The role of the time and place neurons was discussed earlier, and especially how
they encode memories, together with multisensory input. But there is a twist: to
encode a memory, it has to be worth it! Relatability and receptivity are the key filters.
An event we cannot relate to simply slips over our memory and ends up in the "to be
sorted later pile". If we are not receptive, the event will not even be perceived—as
simple as that. A bit like in option #1, when we do not even see Kaïros.

Let's take a typical Murphy's law case: it is raining, you are exhausted after a day
of work, you are loaded with grocery bags, and rushing to your car. Here comes a
little guy, cute even. He would like to get your opinion or your signature for

© Vlad Kolarov, for DervalResearch

Fig. 5.10 Is it the right moment? No (DervalResearch, printed with permission)

something you really care for, like saving panda bears. Your answer is: "Go away, off with your silly survey!" (Fig. 5.10).

Another time, another place. It is hot, you are waiting in a never-ending queue. Suddenly the same little guy—he probably read this book in the meantime—offers you a refreshment. Whatever he wants to tell you, you are more likely to listen. Then of course, if he did his homework and you are in his target, it is the starting point of a great interaction that could be leading to a sale, as it is relatable and you are receptive (Fig. 5.11).

5.5.1 From Milan to Bollywood: The Roberto Cavalli Case

The Roberto Cavalli brand is bouncing back and expanding from Milan to Bollywood. The founder and now still owner of 10% of the brand, Roberto Cavalli, innovated in the 70's with animal prints and bold colors. Celebrities like Brigitte Bardot, and more recently Paris Hilton, Shakira, Jennifer Lopez, or Christina Aguilera have supported the artist—fascinated by textile printing—in building his empire (PTI, 2012). With 90 Roberto Cavalli stores, 54 Just Cavalli stores—the jeans brand made trendy by Norwegian designer Peter Dundas and owned by the same group commercializing Diesel—Roberto Cavalli Children, with 17 stores, and Cavalli Club, the brand is present in New-York, Cape Town, Shanghai, and Abu Dhabi (CPP-Luxury, 2016) (Fig. 5.12).

Even though it created a fuss launching a sexy bikini featuring a Hindu goddess, the Roberto Cavalli brand is loved by Bollywood stars (Priyadarshi, 2004). Diva

Fig. 5.11 Is it the right moment? Yes (printed with DervalResearch permission)

Fig. 5.12 Cavalli Club at Fairmont Dubai (printed with Cavalli Club permission)

Kanika Kapoor, singer and fashion icon, has been spotted at Roberto Cavalli's show at Milan Fashion Week, while Priyanka Chopra, Sonam, and Aishwarya Rai Bachchan were getting noticed in Cavalli at the Cannes Film Festival, especially with that golden fishtail gown!

Difficult to work in our DervalResearch team based on the 5th floor of the Fairmont Hotel in Dubai (Fig. 5.13). when you have a Cavalli Club—the ultimate lounge and night club in Dubai, with zebra prints and Italian food—a couple of floors down, open until 3am. Cavalli appeals to those fierce women, who dare to wear wild

Fig. 5.13 Fairmont Dubai on Sheikh Zayed road in Dubai (printed with Fairmont permission)

prints, and going big. By expanding the brand territory geographically but also by addressing their personas' children, Cavalli enjoys renewed profits. Also, behind the scenes, Roberto Cavalli is doing great business with the Wait Marketing approach, when for instance stylists advise VIPs in their hotel suites. Their outlet in the Wafi mall in Dubai is smartly inviting itself into their clients' room, at the next door luxury hotel Raffles Dubai.

5.5.2   The Wait Marketing 6Ms: The Jaguar Case

Wait Marketing is about communicating at the right moment at the right place, when people are receptive. The Wait Marketing 6Ms—inspired by Kotler's 5Ms of Advertising (Kotler & Keller, 2012)—propose a systematic framework in order to launch an effective communications campaign, as we will see with the Jaguar success story (Fig. 5.14).

The Wait Marketing 6Ms are:

– Mission. What is the brand trying to achieve with the campaign: a purchase? a trial?
– Means. The available budget and resources for the campaign
– Message. It has to be clear and depends on the persona
– Moment. It is key: when and where is the target persona available?

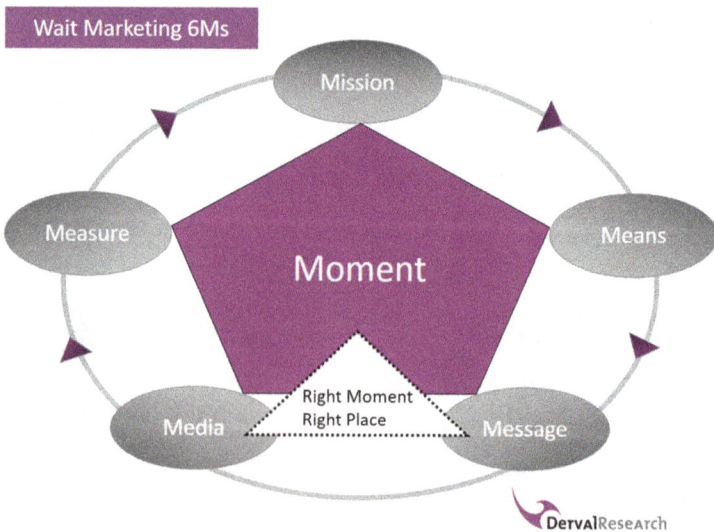

Fig. 5.14 Wait Marketing 6Ms (printed with DervalResearch permission, inspired by Kotler 5Ms of Advertising)

- Media. It will become clear once the best moment and place to interact are identified. If the media does not exist, it can always be invented, and is likely in that case to be extremely affordable. Before being saturated, brand placements in movies, and displays in airport lounges, were effective and very competitive. Now just in a popular Netflix series like "the House of cards", viewers are being exposed to about 100,000 USD of brand placement
- Measurement. It is about making sure to trace the conversion rate by including a distinct # or email, in order to refine the campaign effectiveness over time. 360° campaigns have to be avoided by any means as pushing messages across media makes it impossible to find out what works and what doesn't

Identifying the right luxury persona will not only help develop suitable products and features as we saw in the BMW case, it will also help reach customers and connect with them as we will see in this Jaguar case.

By carefully analyzing the brand personas, Jaguar managed to exceed its sales objectives while saving 30% on its global communication budget.

Vincent, 55 years old, is a member of the board of directors in a large corporation and his revenues are over 200,000 USD per year. He prefers the champagne brand Dom Perignon and he is very much into prestige and individualism. Fashion is a key factor in his purchasing decisions. In between two meetings, he has a look at websites like the Financial Times to check the evolution of his stock options.

Like many deciders, Vincent is very difficult to reach. So Jaguar decided to interact with him over the Internet by posting Jaguar banners on the financial websites he is visiting. The click-through rate was a never-seen-before-on-financial-websites 45%, which means 10 times more than the average performance on the Internet. And the sales followed (Table 5.2).

Based on the success of this first Wait Marketing campaign, Jaguar launched the "Good to be bad" campaign, featuring the new Jaguar F-type coupé and stylish British villain Tom Hiddleston, also known as Loki in the superhero movie Thor. Kicked-off during the 2014 Superbowl, the campaign had a long-lasting effect on consumers' engagement with the brand—some of the ads were even banned because they showcased too-fast driving, adding to the #goodtobebad buzz (Sweney, 2014).

Table 5.2 Wait Marketing 6Ms for Jaguar Villain campaign (DervalResearch, printed with permission, data: Smith, 2017; Sweney, 2014)

Mission	Promote the new Jaguar F coupé and increase Jaguar sales
Means	A great concept and 25 million USD (the price of 400 Jaguar)
Message	"It's good to be bad" message featuring British villains like Tom Hiddleston playing Loki in Thor
Moment	When target consumers are receptive during the Super Bowl break or commuting along the NewYork Metro
Media	Short movies and metro wrap
Measurement	Sales of the Jaguar Land Rover group in the luxury segment increased by 77% year to year, out-performing other luxury car brands

Consumers are more receptive during the Superbowl break because it is a live match and they can understand players need a break, plus they are having fun watching with friends. As opposed to advertising interrupting a movie, as it is quite obvious the actors do not need a break. On top of it, the creative director of Jaguar agency Spark 44 made sure the sound of the V8 engine was very noticeable in the ad. Using the best of Wait Marketing techniques and appealing to non-vibrators, Jaguar implemented the "Good to be bad" campaign at critical moments, like the Superbowl and in strategic places, like New York City where the brand wrapped trains with teasing visuals (Table 5.2). A few happy candidates were proposed a test drive on an F1 racing track via the Jaguar Villain Academy. Again the sales followed as Jaguar Land Rover sales in the luxury models segment increased by 77% year to year (Smith, 2017; Trefis, 2016).

Knowing more about their personas' activities, agenda, and preferences enabled Jaguar to save 30% on their global advertising budget and connect deeply with existing and new customers.

5.5.3 Catch Me If You Can: The Shanghai Tang Case

Another smart aspect of Wait Marketing is to be available on the road of the customer, and Shanghai Tang built its success in travel retail. Shanghai Tang is one of the many brands that used to belong to Richemont luxury group, together with prestigious brands like Cartier, Montblanc, Piaget, Chloe, and IWC to name a few.

Shanghai Tang managed to operate a successful change in its brand codes, modernizing the assortment and style, and attracting more and more local and fashionable customers. With a network of 44 boutiques worldwide, Shanghai Tang, established in 1994, was the first luxury lifestyle brand emerging from China and opened the way for other local designers.

Shanghai Tang considers every consumer touch point as being an important part of the strategy, and the approach and image are therefore very consistent across the shops and countries. Shanghai Tang constantly adapts the assortment, visual merchandising, and packaging to three types of outlets: mansion, travel retail, and fashion boutiques. Their General Manager for China welcomed us to the mansion boutique, where the brand offers a flat experience with a wider scope that includes home furniture, bedding, and a wedding collection. In travel retail outlets, located in airports and luxury hotels, they focus on fast moving seasonal products. In the Xintiandi boutique in Shanghai for instance you will find fast moving products and gifts, like polos and scarves. In the fashion boutiques, usually inside malls, they focus more on fashion goods. A visual merchandising kit is developed for every season and type of boutique (Derval, 2016).

In terms of clothing, men come and they buy five items, women come and buy one or two items—but they come more regularly. Men's fashion is a pretty fast-growing segment. Men are more fashion sensitive and also want to look good.

Shanghai Tang is known for its Mandarin collar and has even created a club called "the Mandarin collar society", gathering opinion leaders from Shanghai, Beijing, Hong Kong, the U.S., and the U.K., who think that ties symbolize being tied. "We promote the use of the Mandarin collar to embrace not only a new sense of fashion, but a new sense of self", shares the General Manager. Children's fashion accounts for about 4% of the sales.

In terms of accessories, men also tend to do more corporate gifting: they buy leather goods, for instance. Women buy accessories like wallets, earrings, and jewelry.

When the new top management joined, the group was reorganizing the consumer targets, and decided to focus on Chinese consumers because of the growth potential. Also Shanghai Tang's concept was by essence naturally appealing to foreigners. In Hong Kong, for instance, most shoppers are expats. The brand is doing very well in Singapore too as customers like the idea of brand codes being a mix between Chinese and other influences.

We had the chance to work with Richemont on identifying the local personas in China in order to design more targeted and appealing collections. The shift was a success, with 50% local customers against only 30% before the revamping. Several years ago, the Shanghai Tang chairman felt a shift was needed and wanted to turn this travel retail success into a leading fashion brand as well. The top management operated a full make-over, reviewing their pricing, retail, product, and communications strategy. When the brand was created 20 years ago, there were a lot of Chinese traditional elements like dragons and phoenix going on everywhere. Now the "authentic" collection represents less than 10% of the sales, driven by women's ready-to-wear, seasonal products, and best-sellers like silk scarves.

Shanghai Tang came up with the concept of China Fashion Chic and took on the role as a leader in Chinese fashion. The brand invited rising Chinese talents to collaborate, and offered them a global presence, thanks to 44 boutiques all over the world. The activities under this long-term vision of China Fashion Chic include: a platform working with designers, capsule collections every year, design forums, design competitions, collaborations with the Chinese fashion week, and international fashion weeks. The collaboration with designers like Masha Ma and Jacky Tsai, and the fashion show featuring Nicole Kidman impressed Shanghai Tang customers and they could perceive the positive change operated by their favorite brand.

To reach a broader target audience, Shanghai Tang partnered for instance with Nespresso and they co-developed a special edition of the popular Citiz model for the Year of the Dragon with a stunning dragon illustration on it. The coffee machine, which was an instant hit, was accompanied by a Chinese lacquered box to display the Nespresso capsules. The brand collaborated also with Moleskine, releasing Chinese horoscope agendas for the Year of the Horse, or the Monkey, and a fun fengshui dairy telling you whether today is an auspicious day for hiking or surgery.

Shanghai Tang started making the most of luxury shoppers receptivity and is now cultivating its relatability, being one of the pioneering Asian inspired luxury fashion brand.

5.6   Take-Aways

Territory

- Luxury shoppers willing to expand their territory can relate to family businesses and their way of operating
- Storytelling is particularly strong when it involves heritage and filiation—strong family values
- Conquering Mars (or Jupiter) is the next big thing for luxury shoppers into expanding their territory, hence the admiration for Elon Musk
- International presence helps build success at home

Sense of Time

- Expertise is a great source of innovation and is what keeps luxury brands alive
- Demonstrating the craftsmanship is an ideal way to showcase expertise and engage customers
- Training is key for teams in contact with customers
- Associating luxury brands with memorable events like the Fashion Week has a very positive effect, as seen with DS

Wait Marketing

- Targeting the right luxury shoppers is not enough, brands have to do it at the right place and at the right moment
- Brands can save on their advertising budget and increase their sales with Wait Marketing, by engaging luxury shoppers when they are more available and receptive
- Travel retail combines Wait Marketing and luxury shoppers' need to travel and is therefore a perfect distribution or promotion channel for luxury brands
- The best way to reach and interact with luxury shoppers is to identify the times they are available using the Wait Marketing approach, following the inspiring DS and Cavalli examples

References

Bertelson, P., & Aschersleben, G. (1998). Automatic visual bias of perceived auditory location. *Psychonomic Bulletin & Review, 5*(3), 482–489.

Bonnefont, Y. (2017, February 9). *Interview by Diana Derval*. Amsterdam: DervalResearch.

Bose, M., Burns, A. C., & Garretson Folse, J. A. (2013). "My fifty shoes are all different!" Exploring, defining, and characterizing acquisitive buying. *Psychology & Marketing, 30*(7), 614–631.

Brooks, C. M. (1962). The concept of humoral control of body function and its significance to the development of physiology. *Humors, Hormones, and Neurosecretions, 17*, 1.

Burgess, N. (2014). The 2014 Nobel Prize in physiology or medicine: A spatial model for cognitive neuroscience. *Neuron, 84*(6), 1120–1125.

Carcano, L., Corbetta, G., & Minichilli, A. (2011). Why luxury firms are often family firms? Family identity, symbolic capital and value creation in luxury-related industries/¿ Por qué las compañías del sector del lujo suelen ser empresas familiares? Identidad familiar, capital simbólico y creación de valor en la industria del lujo. *Universia Business Review, 32*, 40.

CPP-Luxury. (2016, January 20). *Roberto Cavalli Junior boutiques open in Abu Dhabi and Shanghai*. CPP-Luxury.com. Retrieved from http://www.cpp-luxury.com/roberto-cavalli-junior-boutiques-open-in-abu-dhabi-and-shanghai/

Derval, D. (2007). *Wait Marketing. Communiquer au bon moment, au bon endroit*. Editions Eyrolles.

Derval, D. (2010). *The right sensory mix: Targeting consumer product development scientifically*. New York: Springer.

Derval, D. (2016). *Luxury brand marketing* 奢侈品品牌营销:创建·实施·案例. Shanghai: Donghua University Publishing.

Derval, D., & Bremer, J. (2012). *Hormones, talent, and career: Unlock your Hormonal Quotient®*. Berlin: Springer Science & Business Media.

Dreber, A., Rand, D. G., Garcia, J. R., Wernerfelt, N., Lum, J. K., & Zeckhauser, R. (2010). *Dopamine and risk preferences in different domains* (Harvard Kennedy School Faculty Research Working Paper Series).

Eichenbaum, H. (2014). Time cells in the hippocampus: A new dimension for mapping memories. *Nature Reviews Neuroscience, 15*(11), 732–744.

GCI. (2013, August 26). *Luxury buying behaviors, brand perceptions analyzed by Euromonitor*. Global Cosmetic Industry. Retrieved from http://www.gcimagazine.com/marketstrends/regions/bric/Luxury-Buying-Behaviors-Brand-Perceptions-Analyzed-by-Euromonitor-221199591.html

Gohel, K. (2014, May 1). *L'Oréal launches Cacharel travel-retail gifting initiative*. Duty Free News International (DFNI). http://www.dfnionline.com/latest-news/retail/loreal-launches-cacharel-travel-retail-gifting-initiative-01-05-2014/

Jafarpour, A., & Spiers, H. (2017). Familiarity expands space and contracts time. *Hippocampus, 27*(1), 12–16.

Kotler, P., & Keller, K. L. (2012). *Marketing management* (15th ed.). Upper Saddle River: Pearson Education Ltd.

Laniado, E. A. (2015, October 14). *Understanding seasonality in diamond sales*. Retrieved from http://www.ehudlaniado.com/home/index.php/news/entry/understanding-seasonality-in-dia mond-sales

Le Baud, P. (1484, XV). *Compilation des chroniques et histoires de Bretagne*. Paris: BnF, département des Manuscrits, Français 8266 fol. 281.

Miller, D., & Le Breton-Miller, I. (2005). *Managing for the long run: Lessons in competitive advantage from great family businesses*. Boston: Harvard Business Press.

Mohammed bin Rashid Al Maktoum HH Sheikh. (2014, October 20). *The UAE Space Agency and the Emirates Institution for Advanced Science and Technology will collaborate to build the Mars probe*. Retrieved from https://mobile.twitter.com/hhshkmohd/status/524142526194659328

Neville, S. (2013, August 4). Wait in lounge: How Heathrow cashes in on wealthy time-poor travellers. *The Guardian*. Retrieved from https://www.theguardian.com/business/2013/aug/04/heathrow-retail-outlets-designer-brands

Patel, A. D., & Iversen, J. R. (2014). The evolutionary neuroscience of musical beat perception: The action simulation for auditory prediction (ASAP) hypothesis. *Frontiers in Systems Neuroscience, 8*, 57.

Peterson-Withorn, C. (2015, April 20). New report reveals the 500 largest family-owned companies in the world. *Forbes*. Retrieved from https://www.forbes.com/sites/chasewithorn/2015/04/20/new-report-reveals-the-500-largest-family-owned-companies-in-the-world/#2a617663602b

Philip, C. (2015). *Interview by Diana Derval*. Amsterdam: DervalResearch.

Priyadarshi, R. (2004, June 9). Harrods apology over Hindu bikinis. *BBC News*. Retrieved from http://news.bbc.co.uk/2/hi/south_asia/3790315.stm

Pryke, S. R., Andersson, S., Lawes, M. J., & Piper, S. E. (2002). Carotenoid status signaling in captive and wild red-collared widowbirds: Independent effects of badge size and color. *Behavioral Ecology, 13*(5), 622–631.

PTI. (2012, December 9). Roberto Cavalli: About 40 years ago, India was wild and poor. *Daily Bhaskar.com*. Retrieved from http://daily.bhaskar.com/news/ENT-roberto-cavalli-40-years-ago-india-was-wild-and-poor-4105843-NOR.html

Quinn, K., & Bausenhart, K. (2014). The multimodal kappa effect: Context-dependence of sensory dominance in time perception. *Procedia-Social and Behavioral Sciences, 126*, 170–171.

Refinetti, R. (2016). *Circadian physiology*. Boca Raton, FL: CRC Press.

Reinberg, A., Lagoguey, M., Chauffournier, J. M., & Cesselin, F. (1975). Circannual and circadian rhythms in plasma testosterone in five healthy young Parisian males. *Acta Endocrinologica, 80*(4), 732–734.

Romei, V., De Haas, B., Mok, R. M., & Driver, J. (2011). Auditory stimulus timing influences perceived duration of co-occurring visual stimuli. *Frontiers in psychology, 2*, 215.

Russo, N., Foxe, J. J., Brandwein, A. B., Altschuler, T., Gomes, H., & Molholm, S. (2010). Multisensory processing in children with autism: High-density electrical mapping of auditory–somatosensory integration. *Autism Research, 3*(5), 253–267.

Setchell, J. M., & Jean Wickings, E. (2005). Dominance, status signals and coloration in male mandrills (*Mandrillus sphinx*). *Ethology, 111*(1), 25–50.

Smith, G. (2017, January 9). Jaguar Land Rover had its best year ever in 2016. *Fortune Magazine*.

Sweney, M. (2014, July 16). Jaguar 'villain' ad banned for encouraging irresponsible driving. *The Guardian*. Retrieved from https://www.theguardian.com/media/2014/jul/16/jaguar-ad-tom-hiddleston-banned-youtube

Tannenbaum, G. S., & Ling, N. (1984). The interrelationship of growth hormone (GH)-releasing factor and somatostatin in generation of the ultradian rhythm of GH secretion. *Endocrinology, 115*(5), 1952–1957.

Thompson, C. W., & Moore, M. C. (1991). Throat colour reliably signals status in male tree lizards, *Urosaurus ornatus*. *Animal Behaviour, 42*(5), 745–753.

Tolle, V., Bassant, M. H., Zizzari, P., Poindessous-Jazat, F., Tomasetto, C., Epelbaum, J., & Bluet-Pajot, M. T. (2002). Ultradian rhythmicity of ghrelin secretion in relation with GH, feeding behavior, and sleep-wake patterns in rats. *Endocrinology, 143*(4), 1353–1361.

Trefis Team. (2016, September 27). Tata's Jaguar Land Rover continues to accelerate sales by more than its competitors in 2016. *Forbes*. Retrieved from http://www.forbes.com/sites/greatspeculations/2016/09/27/tatas-jaguar-land-rover-continues-to-accelerate-sales-by-more-than-its-competitors-in-2016/#69ec2614798e

Viblanc, V. A., Gineste, B., Stier, A., Robin, J. P., & Groscolas, R. (2014). Stress hormones in relation to breeding status and territory location in colonial king penguin: A role for social density? *Oecologia, 175*(3), 763–772.

Wittmann, M. (2013). The inner sense of time: How the brain creates a representation of duration. *Nature Reviews Neuroscience, 14*(3), 217–223.

Wolch, J., & Dear, M. (Eds.). (2014). *The power of geography (RLE social & cultural geography): How territory shapes social life*. New York: Routledge.

Zheng, Z., & Cohn, M. J. (2011). Developmental basis of sexually dimorphic digit ratios. *Proceedings of the National Academy of Sciences, 108*(39), 16289–16294.

Building Iconic Brands

6

"Classics never make a comeback. They wait for that perfect
moment to take the spotlight from overdone, tired trends".
Tabatha Coffey, celebrity hairstylist and TV personality
(Booth, 2016)

Luxury is driven by iconic brands and classics—think of Chanel N°5 or the Birkin bag. We will see in this chapter how opinion leaders and innovations can meet to create iconic luxury products and memorable experiences.

6.1   Introduction

In this chapter, we decode how to create luxury and fashion icons in Sect. 6.2 with the Maison de Couture case, and answer the listed questions:

- How to start long-lasting trends?
- How to create iconic brands?
- Why do we follow celebrities?
- How to become an opinion leader?
- Why are some people obsessed with Instagram?
- How to recruit the right brand ambassadors?

We discover in Sect. 6.3 that imitating successful individuals is a biochemical tendency you cannot control.

In Sect. 6.4, we explore our chemo-senses and especially the sense of smell, and have a glimpse at Chanel N°5 and Taylor Swift stories.

© Springer International Publishing AG, part of Springer Nature 2018   137
D. Derval, *Designing Luxury Brands*, Management for Professionals,
https://doi.org/10.1007/978-3-319-71557-5_6

In Sect. 6.5, we see how the Influencers' Map can help find the right brand ambassador for each persona through the Montblanc case.

Main take-aways are grouped in Sect. 6.6.

6.2   How to Build an Iconic Brand? The Maison de Couture Case

A brand is just a name. A name you trust because of its past, current, and future great products—its classics. Let's review with the Maison de Couture case luxury brand do's and don'ts regarding brand management and innovation.

6.2.1   Reinventing the Brand

It was one of our first meetings related to media planning. We arrived very excited at the headquarters of this prestigious Maison de Couture in the chic area of Paris. We had a meeting with the VP for Communications. Her assistant welcomed us, super friendly—she was the one who heard of us and set the appointment with her boss. We just had the time to chit-chat a bit and here she arrived all in black, with a style I would qualify as elegant austerity, for what it means. I clearly remember her perfume, very strong and chemical. She sat down and started explaining: "We are a very successful brand as you know, and we do every year a media plan including billboards, magazine ads, to cite a few, and as we are launching the new fragrance, we would like to bring some new perspectives to the teams". I nodded and answered "Sure, innovation is our field. We could start analyzing which media are working better for instance". She replied swiftly "I don't know". I added, "It might be handy to see what could be changed and improved". She stopped me "Change? But we do not want to change anything!" I was getting a bit confused here, "But I thought you wanted to innovate". She confirmed "Of course, but we do not want to change anything as it has been working so far".

I think being asked to help teams "innovate like before" was the most challenging mission we would ever receive.

Successful brands, I would say, especially successful brands, had to reinvent themselves, think of Chanel or Swarovski. The brand codes we studied earlier helped them stay true to themselves and to adapt in the right direction. Hermès was originally designing horse equipment and travel luggage, hence their carriage logo. If you visit their website today, you will still see a strong equestrian section, but the travel luggage offerings are brought to life in little illustrations involving private jets. Because times change.

6.2.2   Launching a New Classic

Classics are iconic products that personify and often survive the designers and the brands. They are also what is needed for luxury brands in difficulty in order to bounce back—think of Swarovski with its jewelry line, dressing celebrities, embellishing hotels, and "ensparkling" designers' creations. At launch, classics were just trends like the others; the difference is they responded to a deep physiological need in a way no other product did or could. Classics are, in a way, sustainable trends. Black is a great example, as we saw in previous chapters: Some shoppers just love black so that a trend based on black will be an instant success and last longer than a season. On the other hand, a trend based on green will be limited to only a couple of designers and fashionistas who can deal with that color and just be ignored by most shoppers. Trends are proposed by designers but consumers can take them or leave them. So understanding their sensory preferences and motivations is key, and social media can help.

So I was eager to ask "Can we smell the new fragrance?" This would help identify the right persona. "The scent itself does not matter as we have the best brand ambassador", stressed the VP, dropping a very popular name indeed. I was not familiar with fragrances at that time, but it sounded odd that just posters combining a celebrity with a powerful brand would do.

6.2.3 From Billboards to Instagram

Major luxury brands are active on social media. One of the most popular ones, Burberry, gained respect and admiration from many followers because of their daring online media strategy.

Which celebrity is wearing the brand is probably more important than the billboard campaign itself, widely replaced by Instagram posts. To be successful a luxury brand needs to dress and equip the *It girl* (the trendy girl). Yesterday Grace Kelly, today Cara Delevingne. And no, dear lady in black from the past, you cannot "innovate like before". Ethically I cannot sell some pricey untargeted 360° advertising campaign involving all media at the same time so you do not know what works in the end—particularly not once you know about the Wait Marketing approach we studied in previous chapters! Unsurprisingly, the new fragrance did not become a classic and sometime later the brand got eventually bought by a larger group and we were able to collaborate with more open-minded and pragmatic teams, grasping better the role of social media and brand ambassadors. Social media enables a greater interaction and we will see that some brands use it not just to push their news, but also to observe their target consumers.

Brand ambassadors are key in the luxury industry. We will see how fashion designers like Pan Hong become true opinion leaders among the wealthy and powerful. We will also see how brands like Montblanc make the most of brand ambassadors, brand placement, and landmarks—the iconic places—to target and engage their customers. The influencers framework will help us structure a brand ambassador strategy.

Before all that, the fundamental question is why do people follow celebrities? The associated question, as useful, would be how to select the right brand ambassador? Also why did the new fragrance not become a classic?

6.3   The Biochemistry of Imitating Successful Individuals

Would you be surprised to hear that our urge (or not) to check celebrities' Instagram and imitate them (or not) is physiological? We will see in this section where this fascination for famous people comes from and how to channelize it to promote a luxury brand. Also we will realize that the type of celebrities we follow is not random and sticks to a very logical pattern.

6.3.1   Following Celebrities

Without noticing it, we spend our time copying successful individuals. At the hairdresser's when asking for this Kate Perry haircut and color, at the nail salon, to get the same claws as Lady Gaga, or when modifying our voice to make it more sexy like a Marylin Monroe or Paris Hilton, before hers started "breaking". Men are also concerned of course, when trimming the beard like Sheikh Mohammed, or multiplying tattoos like Adam Levine.

The way we are attracted to certain celebrities is physiological and chemical. Both men and women are influenced by testosterone and estrogen, while in the womb. The level of exposure to each hormone shapes the way we think, we look, and behave. Based on measurements performed on thousands of people in over 25 countries, DervalResearch identified 4 types of men and 4 types of women. As we studied in the section on gender polymorphism in Chap. 5, you can be more influenced by estrogen like Victoria Beckham and Brad Pitt, more influenced by testosterone like Marilyn Monroe and Donald Trump, very influenced by testosterone like Meryl Streep and Steve Jobs, or balanced (which means equally influenced by estrogen and testosterone) like Madonna, George Clooney, and Gary— our BMW persona in Chap. 1. And we tend to look up to and follow celebrities on this basis, as it is easier to connect with the "same type of beast" or with our "dream beast".

Checking on these celebrities regularly, via Instagram or people magazines, is a way to benchmark and to keep an eye on the prize. These rituals themselves seem— icing on the cake—to procure a feeling of satiety in case of uncertainty (Woody & Szechtman, 2006).

6.3.2   The "Chameleon Effect"

People who are very social, and therefore score high on cognitive empathy, imitate others unconsciously, adopting the same posture, the same gesture like rubbing the nose, and even the same face expression, making themselves look more familiar and engaging (Chartrand & Bargh, 1999). This "chameleon effect" can be linked to findings on the "mirror neurons", a physiological mechanism that makes us imitate our peers (Rizzolatti & Craighero, 2004).

I was first surprised when I read about it as I couldn't relate to this imitation process at all, and research confirmed that folks like me with autism are very different beasts

(Heyes, 2001). For autistic people—who do not care about what others think (this can lead to pure genius or interstellar crap), do not envy others, do not benchmark with others (no, not even checking other girls' rears), and like to be different and unique—imitating is a big no-no. Note that autism has a wide spectrum and most engineers, to cite only this occupation, might be impacted so we could represent a reasonable portion of the society, in general less sensitive to brands and luxury, as we have less the urge to fit in and compete. If attracted to luxury, it would be more for the performance of the actual products than for a given brand or brand ambassador.

6.3.3 The Habits of Successful People

The hope is that imitating the most successful individuals might lead us to become successful as well (Price, Brown, & Curry, 2007). Hence the number of copies sold of the book "The Seven Habits of Successful People" and the success of biographies of leading CEOs and billionaires. Also within a given ethnic group, people tend to interact more with people sharing similar biomarkers such as personality or ornaments and learn from the most successful individuals within their subgroup (McElreath, Boyd, & Richerson, 2003).

   Some argue that the true "wisdom of the crowd" is just about copying successful individuals (King, Cheng, Starke, & Myatt, 2012). Some might then also think that Chinese are very wise—in a sense some Chinese firms (and, let's be honest, many non-Chinese firms too!) did at the level of a company what individuals do at their private level. Is Lady Gaga filing a lawsuit in intellectual property each time her style is being copied? In some countries though, the copying is getting a bit scary, like in Thailand, where bloggers take selfies in the same situation and outfits as their stars. King Shah Rukh Khan in his recent movie Fan also depicted a fan willing to resemble a bit too much his idol.

6.4 Luxury Has a Smell

Let's face it, it is much easier to smell like Beyoncé than to look like her. Celebrities are therefore not only endorsing perfumes, as we will see with the Chanel N°5 case, but launching their own fragrances, like Taylor Swift did. The question is not which celebrities have their fragrances but more which celebrities do not have their fragrances yet. We will decode the secrets behind these best-selling perfumes and investigate the sense of smell and its business applications. And hopefully, we will understand what can make or break a fragrance and create a classic.

6.4.1 The Sense of Smell

Scent is definitively a strong olfactive ornament used by human and non-human animals to find food, mates, and discourage competitors and other predators.

6.4.1.1 The Chemosensory Receptors

Even though it is commonly admitted that amphibians—who were equipped with one chemical sense combining smell and taste—gained two separate senses when starting crawling on the ground, based on my observations, I would say that some people must still have very linked scent and taste receptors, so that when their nose is stuffed, they don't taste anything. We will study the variations in scent perception among individuals with the inhaler profiles, the Derval Pyramid of Scents, and the Sephora case.

Various animals use some kind of smelling or tasting techniques. Snakes smell with their tongue, crabs with their antennas, octopuses use the chemoreceptors included in their 1800 suckers, butterflies and flies taste and smell with their feet. Flies have indeed smelly or, should I say, smelling feet so that when they land on your food it is just to taste it. Flies also like to perfume themselves in a dating context. To do so, they walk all over fermented foods and plants, adding a charming yeast and trashcan scent (Ziegler, Berthelot-Grosjean, & Grosjean, 2013).

The human nose is able to detect 10,000 different odorants, which makes us a microsmatic animal as opposed to macrosmatic animals, like octopuses or dogs, that have a more powerful sense of smell. Still, a mother can identify a piece of clothing worn by her baby, which is not so bad. Maternal breasts also have a distinct odor. And for the record, even sperm cells are equipped with chemoreceptors in order to locate the egg. Olfactory receptors are specialized by proteins and function similarly to rods in the retina, that we saw in Chap. 3. The olfactory receptors are activated by odorants and turn the stimuli into nerve impulse. What makes our sniffing power is the number of receptors and also the binding potential—defined as how strong they bind—with the odorant, which is called the affinity. When we sniff a scent, we inhale odorant molecules through our nasal cavity towards our chemoreceptors (Bittel, 2014).

We also literally smell or exhale what we eat. This explains why celebrity diets are fascinating Instagram or Weibo followers and are even featured in magazines like Harper's Bazaar. If colleagues ask you what you ate the day before, they might be after your job and trying to crack the secret of your success at work (Santos, 2015).

As we saw in Chap. 4 about dating, people eating more carotenoids also have a smell perceived as more pleasant and attractive. While conducting this research on popular scents, we couldn't believe that some people do love the smell of gasoline—the same smell that would give me nausea. Pushing further, we found out that people welcome more or less well scents according to what we named the Derval Pyramid of Scents (Fig. 6.1).

6.4.1.2 The Derval Pyramid of Scents

Most people can cope with the smell of air—when not too polluted—with the smell of food, and mates (Fig. 6.1). At that stage already some disparities might appear. According to broader research conducted in North America and leading to forbidding scents in public settings, 40% of people are irritated by floral, alcohol-based, and chemical scents (Derval, 2010). People can react to real flowers and even more so to perfumes—which in the best case contain alcohol and in the common case

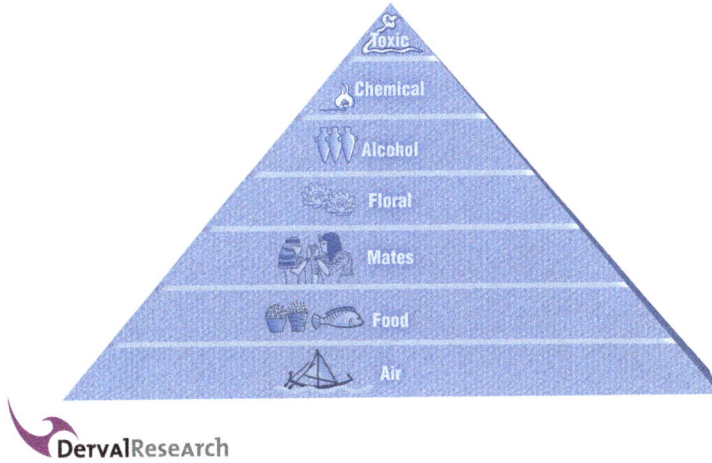

Fig. 6.1 Derval Pyramid of Scents (printed with DervalResearch permission)

chemicals—with various health issues like headaches, asthma, nausea, dizziness, and sinusitis. Food for thought before spreading fragrances in shops, showrooms, and resorts.

The inhaler profiles will help identify target luxury shoppers' sensitivity towards scents and adapt the brand experience and products.

6.4.1.3 Inhaler Profiles

Variations in scent perception are based on physiological differences that are easy to measure and to compare between personas. It is important to understand the inhaler profiles, as we will see with this Sephora example, to be able to select the right assortment, implement the right visual merchandising, and design the right communication campaigns.

Cécilia has a date tonight after the office and she runs to Sephora in Paris for an ultimate beauty consultation and to refill her favorite perfume: Dior J'adore.

Ivy has a date tonight after the office and she runs to Sephora in Shanghai for an ultimate beauty consultation and to refill her favorite perfume: Flower by Kenzo. Remember we discussed Kenzo packaging in Chap. 3.

Sephora—part of the LVMH group—counts over 1700 points of sale across 30 countries and is very active online, selling both luxury brands and its own labels. Always ahead of time, Sephora even created in San Francisco a life-size replica of a Sephora store to test and make sure that new and fancy technologies like augmented reality will actually help improve the brand and shopping experience. The beauty retailer had to adapt to local markets and perception in order to grow and succeed.

At Sephora, Dior J'adore is the best-selling perfume in France with notes of violet, rose, plum, and vanilla; while Flower by Kenzo is the best-selling perfume in China with notes of hawthorn, violet, and vanilla so Cécilia would be a typical French persona and Ivy a typical Chinese persona. Both perfumes have violet and

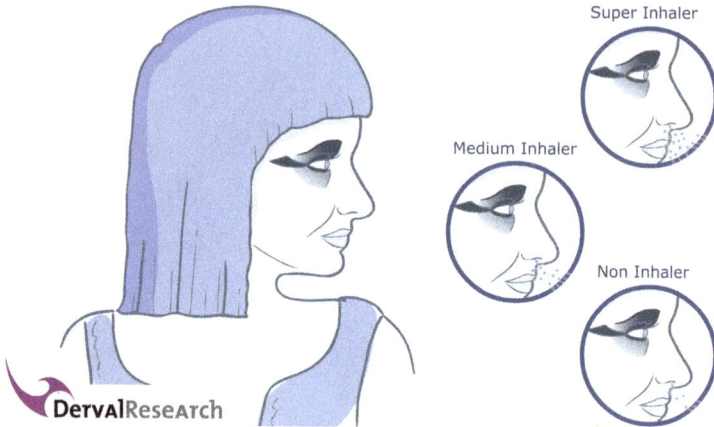

Fig. 6.2 Inhaler profiles (printed with DervalResearch permission)

vanilla as well as musk in common but Flower is more subtle than J'adore. Consumers like Cécilia enjoy that type of intense floral scent and are less sensitive to alcohol.

The best way to compare Cécilia and Ivy is to identify their inhaler profiles. The idea is to identify for a given product—here a perfume, but it could also be a luxury car, a yacht, a Michelin star dinner, or a fashionable dress—the criteria that are critical for the persona when purchasing and see which of these criteria are the most important for each persona.

In our example of a perfume, if both like a floral (violet), fruity (vanilla), and gender-typed scent (musk), their threshold for each criterion seems to vary especially on the floral and musk, where Cécilia is looking for a more intense scent. The OR5AN receptor for instance recognizes musk and might be involved in our attraction to fragrances like Dior J'Adore and Flower by Kenzo (Sebastian & Puranik, 2016).

Based on DervalResearch observations and measurements, we can split individuals into three groups (Derval, 2010):

- Super-inhalers, who are very sensitive to chemicals and prefer a fruity to a floral scent, with little musk
- Medium-inhalers, who like subtle floral scents, not too chemical, like Ivy
- Non-inhalers, who like strong chemical and musky scents, and do not mind the scent of gasoline or Chanel N°5, like Cécilia, our lady in black, or Cleopatra known for burning incense the whole day and being very resistant to chemicals (Fig. 6.2)

The sensitivity of the chemoreceptors can be tested by inhaling capsaicin—but do not do it at home as it makes you cough and can trigger some other nasty reactions.

Table 6.1 Scent preferences by Inhaler profile (printed with DervalResearch permission)

	Non-inhaler	Medium-inhaler	Super-inhaler
Food	Appreciates spicy notes, chocolate, peppermint, and lemon. Doesn't like fishy odors	Prefers baking bread, vanilla, or coffee. Doesn't like fishy odors	Prefers vanilla, and orange notes, or steak smell
Mates	Likes musky notes	Sensitive to musky notes	Very sensitive to musky odors
Baby powder	Doesn't identify easily	Perceives as floral	Perceives as sweet
Floral	Enjoys lilacs, lavender, and wooden notes	Likes jasmine, fresh grass, and rose. Doesn't like sewer or garbage odor	Floral scents are irritating
Luxury fragrances	Chanel N°5, White Diamond, Dior J'Adore	Flower by Kenzo	Trésor by Lancôme
Alcohol	Likes the smell of alcohol and sometimes petroleum	Doesn't mind alcohol smell	Perceives alcohol smell as irritating
Chemical	Less sensitive to chemicals	Can find some chemicals, and cigarette smoke irritating	Very sensitive to chemicals, and cigarette smoke
Population	25%, mainly men	50%	25%, mainly women
Capsaicin test	Few coughs	15 coughs	35 coughs

Based on 1200 observations and measurements conducted by DervalResearch between May 2009 and February 2010 in over 25 countries

The purpose of the inhaler profiles is also to check if a product matches a persona (Table 6.1). We can use a radar chart (in Excel or so) to do the product/persona match making. First, we need to identify around five criteria that define and differentiate our personas and are critical in their purchasing decision. For Cécilia, lasting, musk, and floral are key. For Ivy, floral is important as is fruity. We evaluate for each persona how critical each criteria is on a scale from 1 to 5, for instance, with 1 being not important at all, and 5 being very important. Then we can do the same exercise for the products and evaluate how Flower by Kenzo rates on each of these criteria and even put the perfume on the radar. It then becomes obvious that Ivy and Flower by Kenzo are a good match!

6.4.2 A Classic Scent: The Chanel N°5 Case

According to Coco Chanel, a woman without perfume has no future. No wonder that among the ultimate luxury classics you find several fragrances including legendary Chanel N°5.

Chanel actually built its international success by associating its brand with icons like Marylin Monroe. When the actress revealed that she was sleeping with just a drop of Chanel N°5 as her nightgown, it made a huge sensation in the media and the sales of the perfume exploded, putting Chanel on the international fashion scene, and the brand never left it since then.

In spite of the competition of hundreds of new fragrances—100 new perfumes have been released since 2010—Chanel N°5 is still among the top 4 most sold women's fragrances worldwide. Making it a true classic (King, 2016). What is the particularity of this legendary Chanel fragrance? If we go back in time, it was the first 100% synthetic fragrance—meaning: completely chemical. And it is true that our observations and tests confirmed that this perfume can act as a female competitors' repeller as super- or medium-inhalers cannot take it.

Coco Chanel, rebel and free spirit, innovated in women's wear, introducing pants and cotton jersey—a fabric solely used for men's underpants so far. Chanel helped women reinvent themselves, and Coco through her modern designs was somehow fighting tradition. Chanel dared to innovate and break the rules. Karl Lagerfeld managed to give Chanel an edge back, when the brand started to appeal to "bourgeoises" only—in other words, big wheels' wives. Nowadays international stars like multi-Grammy-winning producer, rapper, singer, songwriter, and fashion designer Pharrell Williams are seen wearing the Paris brand and incorporating some very classic pieces like the long pearl necklace into their street style. Similarly to French Cognac brands—a strong and pricey French spirited alcohol– finding a second life by being adopted by U.S. rappers.

And Chanel N°5 is forever the symbol of powerful, non-inhaler women.

6.4.3   Celebrity Fragrances: The Taylor Swift Case

Elizabeth Taylor was the first celebrity to launch her own fragrance back in 1991. The floral scent generated over 60 million USD sales in total and ironically triggered asthma in its creator. Unlike the Cleopatra she impersonated, Elizabeth is not a non-inhaler for sure.

Top grossing celebrity fragrances so far include Heat by Beyoncé, 400 million USD, Glow by J-Lo, 300 million USD for our Louboutin ambassador, and around 100 million USD for Curious by Britney Spears and Reb'l Fleur by Rihanna (Finances Online, 2014).

Being a celebrity and launching a perfume helps but it won't do the trick alone, as we saw in the Maison de Couture case. In order to be purchased, re-purchased or used as a gift, a celebrity perfume must have a distinct and consistent scent. Since her collaboration with Elizabeth Arden—paradoxically also commercializing successful scents from awarded menswear designer and NBC TV show Fashion Star jury John Varvato and best-selling ever celebrity fragrance White Diamond from Elizabeth Taylor—Taylor Swift fragrances have been a failure. "Wonderstruck", "Wonderstruck Enchanted", "Taylor by Taylor Swift", and recently introduced "Incredible Things" all had a very limited reach, in spite of the singer's huge fame (Brooke, 2014). Why?

It is most likely that the scents just do not fit the celebrity followers. The latest fragrance is based on the winning recipe of vanilla and musk that is behind Dior

J'adore that we studied earlier but the fruity and sweet notes do not really match with the competitive women following the singer and ready to inhale more chemicals and aldehydes, present in perfumes like Chanel N°5 and White Diamond. On the other hand, the musk and amber is too much for women who like sweet and fruity fragrances (Wahba, 2016). Failing to understand the inhaler profile of the luxury shoppers following a key opinion leader (KOL) can make or break a fragrance. And parfum classics clearly address the expectations of the inhaler profile they target. So in our Maison de Couture case, a strong brand and a famous celebrity helps, but cannot replace the right scent, if a repurchase is expected (iscentyouaday, 2016).

6.4.4 Involving Celebrities: Bienvenue chez Maxim's de Paris

Maxim's de Paris is originally a restaurant in Paris. A legendary restaurant, where famous writers and actresses and powerful people used to have dinner, slightly showing off in this very fashionable location with a "Belle Epoque" decoration. Once, actress Brigitte Bardot shocked people by entering the venue with bare feet, turning the restaurant into the place to be.

The brand became also renowned for its catering and opened new locations around the world. In Beijing, for instance, Maxim's has been running its prestigious restaurant for over 30 years. Owned by Pierre Cardin, the famous fashion designer, the brand decided to use chocolate as its symbol and core product when exporting the concept.

So having a clear core delicacy product is always very useful. It doesn't mean that the brand is not doing any other delicious things from champagne to snacks, savory and sweet. But having a strong core product like chocolate—sometimes difficult to manage in hot countries because it can melt—gave Maxim's an edge. "The brand is very active in retail as well as in B2B with company gifting" shared Maxence de Talagrand, in charge of VIP accounts in China, when he welcomed us in Maxim's de Paris boutique in Shanghai. The brand is expanding fast (Derval, 2016).

Maxence invited us to savor some Maxim's de Paris exquisite chocolate bonbons during the interview (Fig. 6.3). We gained some kilos conducting this research for you, dear reader!

Chocolate is a true crowd pleaser, thanks to its three main addictive components: sugar, caffeine, and theobromine (Zoumas, Kreiser, & Martin, 1980). People who have more μ (mu) opioid receptors respond to proteins and to the theobromine and sugar. People who have more κ (kappa) opioid receptors will fall for mint or caffeine. People who have too many of both receptors are doomed (Derval & Bremer, 2012). So the type of chocolate we prefer is directly linked to our opioid receptors. Anyway, chocolate is supposed to minimize the decline faced with aging, so why resist? (Moreira, Diógenes, de Mendonça, Lunet, & Barros, 2016). In the same way coffee smell can give a boost, chocolate smell—I am talking here about the real smell and not some chemical ersatz—can occasion euphoria or alertness, depending on the type of opioid receptors. These properties make it the perfect gift and luxury delicacy.

Fig. 6.3 Maxim's de Paris chocolates (printed and eaten with Maxim's de Paris permission)

Maxim's, in addition to the boutiques and a presence in selected retail, developed a very successful luxury corporate gifting offering, working with brands like BMW or LV.

6.4.5 Inhaler Profiles: Business Applications

As luxury is all about status-seeking and benchmarking with others, celebrities are important role models brands cannot ignore and this offers many business opportunities:

– Creating capsule collections with celebrities
– Considering the fragrance as a main product, not a goodie, and making sure it is consistent with the celebrity and followers' personality and inhaler profile
– As nutrition has a direct impact on appearance and smell, commercializing celebrities menus would be a sensational new business and also allow followers to share their meals with their favorite stars
– Non-inhalers love the smell of gasoline, incense, perfumes, and more and are looking for strong sensory experiences and activities (clubbing, yachting, spa)
– Neutral sensory environments are difficult to find and super- as well as some medium-inhalers would be willing to pay for the luxury of unscented products and places (bags, shoes, hotels)

6.5   Finding the Right Brand Ambassadors

Depending on the luxury brand, the right brand ambassadors might be a celebrity, the founder of a brand and his filiation, a designer, or actual clients.

6.5.1   Celebrity Endorsement: The Visit Dubai Case

Celebrity endorsement, from Lady Gaga to David Beckham, is widely used by brands. In the USA, one advertisement out of five uses celebrity endorsement. Furthermore, celebrities are now launching their own brands, marketing their image. In India, 10% of the celebrities featured in TV ads are famous cricket players and 87% are Bollywood stars. King actor Shah Rukh Khan has for instance endorsed various brands such as Nokia, Airtel, Fair & Handsom, Videocon, Emami Sona Chandi Chywanprash, Pepsodent, Hyundai, and the City of Dubai. In a video targeting Indian expats and friends, Khan surprises tourists in luxury restaurants and resorts around the city. Dubai airport also streams a promotional video in the shuttle but as their budget was more limited it features a penguin—worth the watch nonetheless.

Three criteria to select the right celebrity have been identified: the credibility, the attractiveness, and the product match-up. Credibility is conveyed by the celebrity expertise or trustworthiness. Attractiveness of the celebrity could also play a role in the purchasing decision. Product match-up is about the congruence between the celebrity and the brand. The higher, the better. So far, no direct link was measured between credibility, attractiveness, match-up and sales, but men tend to better trust men and women tend to better trust women.We will see, with the Montblanc case, which men and women are the best candidates and why.

A new generation of key opinion leaders (KOLs) emerged from social media: bloggers. I was always wondering what bloggers were living on as they spend the whole day doing selfies. It happens to be a quite lucrative activity: An influencer with let's say 1 million+ followers can negotiate a branded post—which is a brand placement on their social media—around 125,000 USD on Youtube and 65,000 USD on Facebook. Every month, more than 200,000 branded posts flourish on various social media. Of course, celebrities known for specialized talents also benefit from this new brand placement frenzy. Recently, football player Cristiano Ronaldo displayed his nice Tag Heuer gift, posting even the brand's catchline to his 240 million followers, in case there was still a doubt on whether it was a branded post or not (Badenhausen, 2017). The most popular social media with over one billion users are YouTube, Facebook, and WhatsApp, closely followed by WeChat, and then with half a billion users or so you have Instagram and more business related LinkedIn (SocialBakers, 2016). We will see later, with the Pan Hong case, how some lucky ones become influencers. Also given these fees, finding a genuine brand ambassador, someone who truly loves your products, might be a good idea.

Proposing appealing products can also save a lot of budget, for instance KOS (King of Sheepskin), a premium designer boots brand keeping UGG on their toes, managed to wow the platinum singer Alexandra Burke, winner of the British singing competition X-Factor. Alexandra is the official brand ambassador of Dolce & Gabbana with a 6-figure contract but has been seen proudly wearing her KOS boots just because she loves them and they come in many fierce, bling, and bold variations. Other fans of the brand, created by the godfather of sheepskin, include Sarah Jessica Parker and Sienna Miller. KOS is growing fast following the footsteps of UGG. The luxury boots brand with an Australian flair was actually created in the USA by a surfer, and the shoes, first adopted by Olympic Games athletes, gained in popularity after being favorited by Oprah Winfrey. Strategic partnerships with Jimmy Choo and Swarovski also helped in creating the now classic UGG boots.

6.5.2   The Influencers' Map Framework: The Montblanc Case

Having strong brand ambassadors is key. The influencers' map will help us identify the right ambassadors for a luxury brand, as we will see with the Montblanc example.

6.5.2.1 Mapping KOLs per Persona

Why did Tag Heuer approach Ronaldo? They might have identified that the football champion's followers were a perfect target for their new mid-entry level watch. Or they might just have thought "Well, he has millions of followers, let's spam them all". Second option is obviously a waste of time and money. Brand ambassadors have to be carefully selected based on the target personas and the influencers' map will help us in this task.

First we need to identify each persona, their favorite hobbies and movies. Which actress or sportsman are they looking up to? Then we try to identify celebrities linked to these universes, to whom personas can relate to and who would nicely represent the brand. If your persona is looking up to powerful female singers and your brand is fierce, Miley Cyrus might be the right choice. If the brand codes are more into timeless and elegant, you might want to go for Céline Dion.

In fine writing, like in the flower business, what really matters is the size. Here it is not about the length of the flower stalk—as we discovered earlier when talking with Monceau Fleurs managers—but the breadth of the fountain pen nib. For instance, Arnold Schwarzenegger uses a Pelican 1000 double broad and Prince Charles a Parker Duofold triple broad!

For some, luxury pens are synonymous with power. In ancient times, deals were negotiated with the sword, then with fountain pens, more recently with a golden iPhone, and tomorrow with the next device you can bring to the meeting room to impress your business contacts—maybe a smart watch? Yes, while this book was being printed, Montblanc announced its Android smart watch Summit, for mountain/social climbers. A luxury pen maker like Montblanc, part of the Richemont group (we mentioned some of their other brands like Cartier and Shanghai Tang in previous chapters), understood this shift and is now focusing on watches in its branding strategy.

The idea of the influencers' map is to identify for each target persona who they look up to. Do they admire and follow actors? Politicians? Men tend to take as role models those who risk their lives: mountain climbers, sailors, extreme sportsmen, astronauts, pilots, and of course super heroes. All super-proprioceptors, of course, like Batman—remember from Chap. 2. No wonder Montblanc selected Chris Hemsworth, the actor playing the role of Thor, god of thunder, in the Avengers movie, to promote their products.

6.5.2.2 A Congruent Ambassador

Montblanc even appointed another super hero, Hugh Jackman, aka Wolverine in X-Men, to be its global brand ambassador. The brand confirms the congruence by stating that Jackman "represents all the attributes of the Montblanc brand: elegant, talented, pioneering, and committed to the arts". Other luxury brands are also surfing on the superheroes wave. Jaguar, remember, featured in a recent Wait Marketing campaign the villain Loki, brother of Thor, and talented actor Tom Hiddleston. Many luxury watch manufacturers target golf players and Formula 1 racers but what if you discover that your personas actually fancy badminton? Then, like Montblanc, you would appoint the Chinese badminton player Lin Dan as your brand ambassador. Lin Dan is recognized by his peers as a "man of style and substance". Being a Montblanc brand ambassador is a role for champions only as it is difficult to impersonate the brand's 100-year heritage, with sublime pieces like the fountain pen Meisterstück. When your personas fancy horse-riding and also secretly everything related to royalty, Charlotte Casiraghi is the ideal brand ambassador. Note that this eminent horse jumper appointed as brand ambassador by Montblanc happens also to be a fashion icon and Princess of Monaco! As a result, she is attending various events ranging from Formula 1 races to charity gala dinners: as many opportunities to represent the Richemont brand (Fig. 6.4).

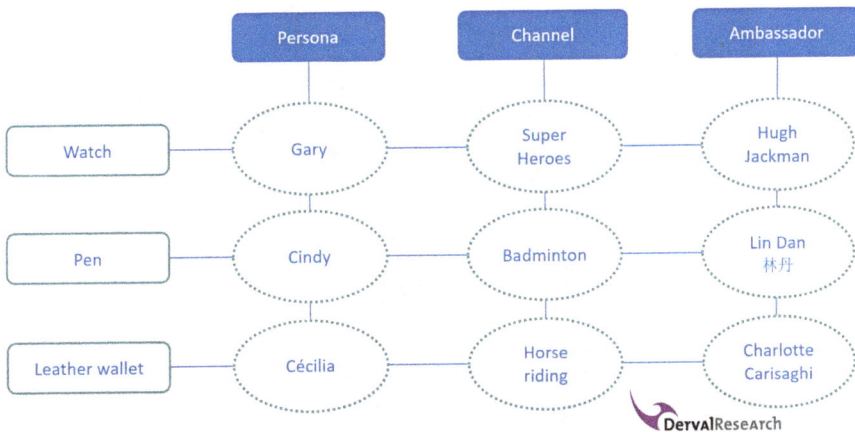

Fig. 6.4 Influencers' Map Montblanc (printed with DervalResearch permission)

6.5.3   Being an Opinion Leader: The PAN's Case

The media call her "the Chinese Coco Chanel". We were welcomed to designer Pan Hong's studio in vibrant Shanghai (Fig. 6.5). Just opening a second studio in Hangzhou, the housewife taught herself design and built the successful tailormade brand PAN's by the power of word-of-mouth only. Her style is contagious and her craftsmanship acclaimed—she recently wowed the fashion scene by creating a fur-looking fabric made of silk. Pan Hong agreed to share some insights on how to promote a luxury brand and become a key opinion leader.

Fig. 6.5 Designer Pan Hong (printed with PAN's permission)

Pan Hong admitted not paying great attention to other designers because she feels that checking their work too much would influence her own inspiration. She gets her inspiration from movies and traveling. And she acts as an ambassador for her own brand. Whether she travels to Tibet or to Paris, she prepares her 10-days clothing for the trip, considering the local culture, weather at destination, and which of her designs would fit best. On top of it, she takes and posts breathtaking pictures. Her VIP customers are all her real-life friends: they enjoy a very good financial situation, like her, and are very keen on traveling but do not really know how to dress appropriately. "Imagine it is very cold in Tibet and you are wearing a Moncler down jacket: you would feel warm but wouldn't fit in the environment", explains Pan Hong. She manages to get her friends interested in her designs, always in a subtle way as it would feel awkward to be too pushy with friends. Versatility is valued by luxury customers and seen as a sign of performance in women. This explains why female celebrities and bloggers—think of Kylie Jenner or Kate Perry—are most followed when they post pictures in various settings with always very appropriate styles and outfits.

6.5.3.1 WeChat Marketing: Selling to VIP Friends

Pan Hong's customers come from different places but mainly from Shanghai, Jiangsu, and Zhejiang provinces—richer neighborhoods. They are from all walks of life ranging from TV hostesses, actresses, and business owners to socialites, or gold and white collar workers. The regular customers used to buy Chanel and Dior but now 90% of their wardrobe is labelled with her brand PAN's—now that's some penetration strategy! They tended to buy good quality and a big logo. When they first visit her design studio, she provides them with styling advice. Then they go out, wearing the clothes, and get validation from friends and relatives. Thus, they build confidence and the brand builds a good reputation. In addition to being a tiny bit more affordable than a Chanel, PAN's clothes are more suitable and make a stronger impression than the Western top-level brands. That is how the brand attracts and retains very loyal customers (Derval, 2016).

Pan Hong will get to know her customers' personality and thanks to their long-lasting relationship, she will provide them with comprehensive advice based on their appearance, occupation, and lifestyle. "My customers are successful in business, so usually nobody dares to give them fashion advice! Although I can't change them into Fan BingBing (famous Chinese actress), I can guarantee that people in their entourage will notice the difference", she shares. Once customers become VIP customers—it would usually take 1 or 2 years—they don't need to come to the design studio anymore: they just have to WeChat Pan Hong what type of event they will attend in the coming week or month, and she will design for them, after confirming the feasibility. It is common in China to use WeChat—I would describe it as an advanced version of WhatsApp—to stay connected with clients, as good clients usually also become friends or the other way around.

6.5.3.2 Putting Customers on the Runway

"My favorite colors are black, white, and grey, and that's why they are PAN's main colors. I randomly add some other colors: for Chinese New Year I will add some red, and if pink is on trend, I will design something pink, but the main colors are black, white, and grey. These are my brand codes", adds Pan Hong, who is a clear monochromat according to our Derval Color Test® from Chap. 3. For members, the designer creates Haute-Couture unique pieces, for regular customers, she adapts existing styles.

"I use natural materials like silk and cotton. I used to have some leather or fur occasionally, but not anymore", she continues. "Some of our PAN's clothes, like this one, look like real fur but in fact it is made out of silk. It is very time consuming to make this kind of clothes. It takes a skilled worker three or more weeks to do just one item by cutting silk pieces one centimeter by one centimeter before putting them on the clothes", highlights the designer. This type of conceptual piece is released in a limited number of five pieces only.

Invited to many parties, Pan Hong prefers to spare her time for her customers. Also, she can only create in a quiet environment. "If I socialized every day, I don't think I would have time to design", she emphasizes. In Shanghai many designers used to do custom made, but gradually they gave up and started solely doing branding. Pan Hong feels optimistic about the rise of the haute-couture in China and believes that people seek something unique in the end. Pan Hong's vision is that "Ten years ago, most business people had a strong passion for leading fashion brands. Nowadays, people and especially the younger generation, like those born after the 80's and 90's, prefer designers' brands. They don't follow the trends blindly nor imitate celebrities, as they believe they are the one and only."

This is why, during her last fashion show, she invited real customers to walk on the runway. Each and every one of them acting as PAN's brand ambassador.

6.5.4   Turning Trends into Classics: Fashion Afternoon-Tea at Raffles Dubai

One of the ten landmarks listed by VisitDubai.com (VisitDubai, 2017) is actually a hotel—The Raffles Dubai. Located right next to the Wafi Mall, in the center of the city, the 19-floor luxury hotel was inspired by the great pyramids of Egypt. The stunning architecture of this pyramid-shaped hotel competes with Burj Khalifa and the Burj Al-Arab towers (Fig. 6.6). When you enter the pyramid, you discover a marble lobby, ottoman-style sofas, an impressive ballroom with Swarovski chandeliers, and while you explore the hotel and the gardens (yes, gardens in Dubai!) you might come across walls with hieroglyph-type carvings—the stones and carvings actually coming directly from Egypt, adding even more authenticity to the venue.

The Royal suite is Turkish-themed with golden accents and the four elements theme appeals to the super-polarotactic guests and blends with the Middle Eastern and Asian flair. You might bump into the masterpieces of famous British fashion illustrator Hatty Pedder and discover Raffles Dubai Fashion Afternoon-tea (Raffles 2013). The incredible sweet and savory pieces look like they just escaped the runway and feature for instance Spanish Manolo Blahnik high heels—with cakes in the form of the classic Hangisi pumps—or the latest designer bags (Fig. 6.7). Austrian Executive Pastry Chef Roland Eitzinger, rightfully nominated for the Best Pastry Chef of the Year awards by the Leaders in F&B (Food and Beverage), managed with his team to replicate wonderful fashion pieces and to turn them into delicious sweets you almost do not dare eating.

Fig. 6.6 Raffles Dubai luxury hotel (printed with Raffles Dubai permission)

Fig. 6.7 Raffles Dubai Afternoon Tea (printed with Raffles Dubai permission)

You can get your pair of Hangisi pink satin shoes at Saks Fifth Avenue but it will cost you 1000 USD or can join the Raffles Dubai afternoon tea and hope to be one of the lucky ladies to win her own pair of Manolo's in a special promotion raffle!

Many luxury venues, whether resorts or malls, suffer from just being a trend, so that once the trend fades, guests go to the latest hot place. In Dubai, locating a venue in or near a highest tower, and even better inside a stunning pyramid is a smart way to turn a trend into a classic. This is the reason why for many luxury brands, location is everything. Combining iconic products and famous landmarks, Raffles Dubai managed to build and sustain its image as a great luxury resort and is the place to be. Ayman Gharib, the trend-setting General Manager of Raffles Dubai, who warmly welcomed us (thanks again for the fantastic sushi lunch at the TOMO Japanese restaurant at Raffles Dubai), shared his vision of luxury: "Luxury is whatever makes you feel special" (Gharib, 2016). Raffles Dubai's major asset is its people, from the service-oriented day and night manager, to the valets with their funky hats and great smiles. It is a whole team taking care of you—and trust me, I am the most demanding guest.

6.6   Take-Aways

Imitating

- People tend to imitate successful individuals
- Followers want to know all the details from the followed person's life as they do not know what led to success, so that in case of doubt they will mimic everything from the diet to the fashion style
- Versatility is valued by luxury customers and seen as a sign of performance in women, hence the popularity of actresses
- Proprioception is valued by luxury shoppers in men, hence the popularity of sportsmen

Sense of Smell

- Variations in the perception of smell are huge and some odorants like musk or chemicals need to be handled carefully depending on the target personas
- Real scents like flowers can trigger negative or positive reactions that will depend on the inhaler profile

Brand Ambassadors

- Nothing can beat a celebrity who really uses the luxury products she/he promotes
- Using brand ambassadors helps create an affinity between the brand and the consumers
- Designers are the first opinion leaders of a brand
- Ideal is to find a brand ambassador for each persona based on their hobbies

References

Badenhausen, K. (2017, June 15). Cristiano Ronaldo produced nearly $1 billion in value for sponsors on social media. *Forbes.* Retrieved from https://www.forbes.com/sites/kurtbadenhausen/2017/06/15/cristiano-ronaldo-produces-nearly-1-billion-in-value-for-sponsors-on-social-media-this-year/

Bittel, J. (2014, November 22). Four weird ways animals sense the world. *National Geographic.* Retrieved from http://news.nationalgeographic.com/news/2014/11/141122-crabs-snakes-smell-taste-nose-science-biology/

Booth, D. (2016, December 26). Motor Mouth: Buick and Volvo are the comeback kids. *Motormouth.* Retrieved from http://driving.ca/buick/auto-news/news/motor-mouth-buick-and-volvo-are-the-comeback-kids

Brooke, E. (2014, October 1). Here's hoping Taylor Swift's new fragrance sells better than the last one. *Fashionista.com.* Retrieved from https://fashionista.com/2014/10/taylor-swift-fragrance

Chartrand, T. L., & Bargh, J. A. (1999). The chameleon effect: The perception – Behavior link and social interaction. *Journal of Personality and Social Psychology, 76*(6), 893.

Derval, D. (2010). *The right sensory mix: Targeting consumer product development scientifically.* New York: Springer.

Derval, D. (2016). *Luxury brand marketing* 奢侈品品牌营销:创建·实施·案例. Shanghai: Donghua University Publishing.

Derval, D., & Bremer, J. (2012). *Hormones, talent, and career: Unlock your Hormonal Quotient®.* Berlin: Springer Science & Business Media.

Finances Online. (2014). 10 most successful celebrity scents: Signature perfumes by Rihanna, Beyoncé and more. *Finances Online.* Retrieved from https://financesonline.com/10-most-successful-celebrity-scents-signature-perfumes-by-rhianna-beyonce-and-more/

Gharib, A. (2016, December 19). *Interview by Diana Derval.* Amsterdam: DervalResearch.

Heyes, C. (2001). Causes and consequences of imitation. *Trends in Cognitive Sciences, 5*(6), 253–261.

iscentyouaday. (2016, January 26). Taylor Swift incredible things. I scent you a day. Retrieved from http://iscentyouaday.com/2016/01/26/taylor-swift-incredible-things/

King, J. (2016, July 22). Perfume marketing based on nostalgia syncs best with consumer behavior. *Luxury Daily.* Retrieved from https://www.luxurydaily.com/perfume-marketing-based-on-nostalgia-syncs-best-with-consumer-trends/

King, A. J., Cheng, L., Starke, S. D., & Myatt, J. P. (2012). Is the true 'wisdom of the crowd' to copy successful individuals? *Biology Letters, 8*(2), 197–200.

McElreath, R., Boyd, R., & Richerson, P. (2003). Shared norms and the evolution of ethnic markers. *Current Anthropology, 44*(1), 122–130.

Moreira, A., Diógenes, M. J., de Mendonça, A., Lunet, N., & Barros, H. (2016). Chocolate consumption is associated with a lower risk of cognitive decline. *Journal of Alzheimer's Disease, 53*(1), 85–93.

Price, M. E., Brown, W. M., & Curry, O. S. (2007). The integrative framework for the behavioural sciences has already been discovered, and it is the adaptationist approach. *Behavioral and Brain Sciences, 30*(1), 39–40.

Raffles. (2013, April 17). Raffles celebrates art for 2013. *Raffles Hotels & Resorts.* Retrieved from http://m.raffles.com/press-room/news/raffles-celebrates-art-for-2013/

Rizzolatti, G., & Craighero, L. (2004). The mirror-neuron system. *Annual Review of Neuroscience, 27,* 169–192.

Santos, S. K. (2015, June 12). 14 best celebrity meals on Instagram. *Harper's Bazaar.* Retrieved from http://www.harpersbazaar.com/culture/travel-dining/g5845/best-instagram-celebrity-meals/

Sebastian, S., & Puranik, N. (2016). Recent concepts about sense of smell, odorant receptors and physiology of olfaction-an insight. *Physiology and Pharmacology, 20*(2), 74–82.

SocialBakers. (2016). Social media overview 2016. *SocialBakers.* Retrieved from http://www.socialbakers.com

VisitDubai. (2017). Top ten places to visit in Dubai. *VisitDubai.* Retrieved from http://www.visitdubai.com

Wahba, P. (2016, February 4). The maker of Taylor Swift perfume says celebrity fragrance sales are falling. *Fortune.* Retrieved from http://fortune.com/2016/02/04/elizabeth-arden-taylor/

Woody, E. Z., & Szechtman, H. (2006). Uncertainty and rituals. *Behavioral and Brain Sciences, 29*(6), 634–635.

Ziegler, A. B., Berthelot-Grosjean, M., & Grosjean, Y. (2013). The smell of love in Drosophila. *Frontiers in Physiology, 4.*

Zoumas, B. L., Kreiser, W. R., & Martin, R. (1980). Theobromine and caffeine content of chocolate products. *Journal of Food Science, 45*(2), 314–316.

Conclusion

Writing this book, I discovered I was a super-vibrator, performance-driven, non-proprioceptor, tetrachromat, non-polarotactic, indexed, super-inhaler with a testosterone Hormonal Quotient® (HQ). What about you? Through the various business cases, interviews, and insights I hope you sensed that luxury is in the end quite simple to decode.

We are all consumers, but first people, and even before that, animals. Our physiological makeup will condition our motivation in life, and taste for luxury. So everything we do, from buying a rose gold iPhone to showing off our holiday pictures on social media is just a result of who we are. This is the reason why luxury shoppers are so unique and at the same time so predictable.

Luxury and fashion is all about competing and mating, and it becomes clear that for some people luxury items are a must-have, not just a nice to have. Luxury can help us impress our peers, date the most successful male or female, become more powerful, and get successful children. No car, outfit, jewelry, high-heel shoes, or handbags are too expensive if they help achieve this goal. Every trick is allowed when it comes to the survival of the fittest, including surgery.

That is why luxury is the center of many people's lives and is a flourishing industry no matter the economical situation. Understanding the mechanisms underlying luxury shoppers' behavior and preferences is the key component of luxury marketing.

I hope the scientific knowledge on luxury shoppers' drivers, perception, and preferences, and the tools we shared—like persona, customer benefits, positioning map, brand codes, Wait Marketing 6Ms, and influencers' map—will help you implement winning strategies to create, revamp, and develop your luxury brand.

© Springer International Publishing AG, part of Springer Nature 2018
D. Derval, *Designing Luxury Brands*, Management for Professionals,
https://doi.org/10.1007/978-3-319-71557-5

About Prof. Diana Derval

Prof. Diana Derval is Chair and Research Director of DervalResearch 代戈, global research firm specializing in shopping behavior, and owner of the luxury fashion accessories brand Derval Paris. Harvard Business Review contributor, inventor of the Hormonal Quotient® (HQ), and of the Derval Color Test® taken by over nine million people, nominated for the 2013 Edison Awards, member of the Society for Behavioral Neuroendocrinology, and author of the books "Wait Marketing", "Hormones, Talent, and Career", and "The Right Sensory Mix" finalist of the Berry-AMA Award for best marketing book 2011—Diana turns fascinating neuroscientific breakthroughs into powerful business frameworks to identify, understand, and predict human traits, motivations, and behavior. She has accelerated the development of Fortune 500 firms including Sephora, Michelin, Sofitel, Philips, and Louis Vuitton. Regularly featured in the media (Fujian TV, Guangxi TV, France Info, Marketing Management, Sidney Herald Tribune, Fashionista) Diana Derval teaches Innovation, Luxury, and Neuromarketing at Sorbonne Business School, IESEG, INSEEC, ISM, ESSEC Paris-Singapore, IFA, Fudan, Donghua, and Jiaotong University. Over 23,000 professionals have enjoyed her inspirational lectures and workshops from Paris to Shanghai.

© Springer International Publishing AG, part of Springer Nature 2018
D. Derval, *Designing Luxury Brands*, Management for Professionals,
https://doi.org/10.1007/978-3-319-71557-5

Books by the Same Author

Derval, D. (2006). *Wait Marketing: Communiquer au bon moment, au bon endroit.* Paris: Eyrolles.

Derval, D. (2010). *The right sensory mix: Targeting consumer product development scientifically.* Heidelberg: Springer.

Derval, D. (2011). *Réussir son étude de marché en 5 jours.* Paris: Eyrolles.

Derval, D. 戴安娜·代尔瓦勒 (2016). *奢侈品品牌营销:创建·实施·案例.* Shanghai: Donghua University Press.

Derval, D., & Bremer, J. (2012). *Hormones, talent, and career*: Unlock your Hormonal Quotient®. Heidelberg: Springer.

About DervalResearch

DervalResearch is the worldwide leading research firm in behavioral neuroendocrinology applied to product development and communication. With predictive segmentation models like the Hormonal Quotient®, and groundbreaking neuromarketing tools like Sensory GeoMaps®, Persona Quest®, Wait Marketing, the Derval Color Test®, and the Sensory Lab®, our team of scientists helps brands to understand consumers, increase their innovation hit rate, and deliver the right products and experience from Paris to Shanghai.

You will find more information at www.derval-research.com

© Springer International Publishing AG, part of Springer Nature 2018
D. Derval, *Designing Luxury Brands*, Management for Professionals,
https://doi.org/10.1007/978-3-319-71557-5

Index

10

10 Corso Como, 117–118

A

Affiliation, 32, 33, 110–111
Affordable luxury, 63–65, 74
Alpha female, 54–55
American Express, 17–18
Assortment, 18, 57, 64, 71, 73, 77–79, 88, 93, 132, 143

B

Bag, 12, 16, 32, 34, 45, 46, 51–55, 63, 64, 66, 71, 88, 106, 110, 111, 126, 137, 148, 154, 159
Bag bitch, 53
Batman, 36, 37, 40, 151
Bayerische Motoren Werke (BMW), 1, 2, 6, 7, 10, 11, 16, 17, 20–24, 34, 39, 66, 131, 140, 148
Benefits framework, 27, 45–48
Biochemistry, 140–141
Blacksocks, 2, 10, 16, 18–20
Bollywood, 12, 30, 32, 70, 128–130, 149
Brand ambassadors, 12, 18, 45, 98, 110, 137, 139, 149–156
Brand codes framework, 43, 96–99
Brand essence, 96–97
Brand naming, 15–16
Brand placement, 131, 139, 149

C

Carotenoid, 81, 85–87, 90, 92, 107, 108, 142
Cara Delevingne, 139
Cars, 1, 4, 5, 7, 11, 15, 21–24, 28–30, 33, 35, 37, 54, 79, 82, 88, 89, 91, 101, 118, 120–124

Celebrities, 6, 32, 34, 54, 79, 93, 97, 99, 110, 128, 137–141, 147–150, 153, 154
Celebrities' endorsement, 149–150
Chameleon effect, 140–141
Chanel N°5, 137, 141, 144–147
Charles Philip, 123–125
Chemosensory receptors, 142
Chinese millionaires, 2–5, 65
Chinese red, 95–96
Christofle, 77, 89, 93–95
Circadian rhythm, 111–112
Classic, 43, 52, 137–139, 141, 145–147, 150, 154–156
Color, 4, 5, 20, 21, 28, 31, 45, 47, 51, 56–66, 70, 72, 74, 75, 78, 79, 81, 82, 84, 85, 88, 91–93, 95, 96, 98, 107, 123, 128, 139, 140, 153, 161, 165
  lenses, 65–66
  profiles, 56–59, 66
Competitors, 6, 17, 45, 47, 51, 53, 54, 66–70, 74, 75, 141, 146
Complications, 12–14
Cones, 40, 58–60, 75, 84, 85, 90–92
Contrarotator, 45–48
Cosmetics, 66, 67, 105–107
Countries, 11–13, 17, 20, 27, 31, 33, 34, 39, 44–46, 57, 58, 60–62, 66, 70, 79, 89, 91, 94–96, 106, 109, 117, 124, 132, 140, 141, 143, 145, 147
Cristiano Ronaldo, 149
Customers' benefits, 27, 28, 44–48

D

Derval Color Test®, 56–59, 64, 72, 85, 91, 154, 161
Derval Paris, 119, 121, 161
Derval Pyramid of Scents, 142–143

© Springer International Publishing AG, part of Springer Nature 2018
D. Derval, *Designing Luxury Brands*, Management for Professionals,
https://doi.org/10.1007/978-3-319-71557-5

DervalResearch, 3, 4, 7–10, 20, 24, 30, 33, 34, 36, 38, 39, 48, 55, 57–59, 61, 62, 64, 68, 71, 73, 83, 85, 88–90, 98, 109, 113–115, 117, 126, 127, 129–131, 140, 143–145, 151, 161, 165
Designer bag, 32, 52–53, 55, 154
Diamond, 5, 34, 65, 68, 77–79, 81, 82, 86–93, 95, 99, 101, 111, 145, 146
Diamonds' 4Cs, 78
Dress, 6, 28, 31, 52, 66, 72, 84–86, 89, 96, 117, 124, 138, 139, 144, 153
Drivers, 13, 15, 23, 27, 32–35, 39, 48, 108, 118, 123, 159
Drone, 30, 35, 36
DS Automobiles, 99
Dubai, 27, 28, 30, 31, 52, 53, 65, 66, 70–74, 128–130, 149–150, 154–156
Dyson, 16, 19, 28, 44–48

E
Electrogastrogram (EGG), 40
Elizabeth Taylor, 146
Elon Musk, 15, 134
Ermenegildo Zegna, 15–16

F
Family, 16, 23, 32, 100, 106, 110, 111, 117, 125, 134
Fan Bingbing, 153
Fashion, 4, 22, 31, 33, 34, 41–44, 52–54, 56, 62, 63, 65, 66, 70, 72, 74, 98–100, 106, 111–113, 118, 119, 121, 123–125, 128, 131–133, 137, 139, 146, 147, 151–154, 156, 159, 161
Fashion accessories, 52, 66, 119, 161
Fashion Week, 42, 56, 63, 111, 119, 121, 124, 125, 129, 133, 134
Female-to-female competition, 51, 53–56, 66, 74
Fragrance, 45, 60, 65, 106, 138, 139, 141, 143–148
Free-riding, 40–41, 43
Functional Magnetic Resonance Imaging (fMRI), 83

G
Gender polymorphism, 56, 66, 105, 107–111, 140
Geographical area, 3, 11, 27, 31–32, 99
Gifting, 46, 55, 65, 67–69, 105–107, 133, 147, 148
Global, 41, 101, 107, 123–125, 131–133, 151

Gold, 2, 4–6, 14, 17, 28, 31, 41, 56, 61, 65, 67, 69, 71, 81, 84–87, 89–91, 93, 101, 116, 153, 159
Golden iPhone, 1–6, 8, 16, 22, 24, 31, 68, 82, 87, 91, 150
Grand Optics, 51, 59, 65, 66, 69–74

H
Haidinger's Brush, 84–86, 89, 93
Harley Davidson, 1, 7, 8, 11–12, 39
Hats, 119, 156
Haute-couture, 62, 65, 119, 124, 154
Healthy, 80–81, 90
Hormonal Quotient® (HQ), 23, 55, 71, 73, 107–110, 159, 161, 165
Hormones, 1, 5–7, 10, 56, 71, 75, 107, 108, 112, 161, 163

I
Iconic brand, 137–156
Imitating, 137, 140–141, 156
India, 11–13, 99, 111, 149
Influencers' map, 138, 150, 151
Infradian rhythm, 111–112
Inhaler profiles, 142–145, 148
Instagram, 66, 69, 137, 139, 140, 142, 149
Internal mechanics, 35–37
iPhone, 1, 2, 4, 6, 8, 9, 15, 16, 19, 22, 24, 31, 67–69, 82, 87, 91, 150, 159

J
Jack Ma, 2, 15, 29
Jaeger-Lecoultre, 1, 7, 10–14, 115
Jaguar, 106, 125, 130–132, 151

K
Kardashians, 56, 93
Karl Lagerfeld, 146
Kate Spade, 51, 52, 58, 63–65, 74
Katy Perry, 54
Key opinion leaders (KOLs), 75, 110, 147, 149–152

L
Lady Gaga, 140, 141, 149
Laws of attraction, 79–81
Light polarization, 82, 83, 86, 93
L'Oréal, 92, 106
Louboutin, 16, 31, 34, 58, 78, 81, 95–96, 101, 117, 146

Louis Vuitton (LV), 53
Luxury brands, 1, 2, 10, 16, 17, 19, 21, 24, 27,
    31, 33, 34, 41, 48, 51, 53, 57, 63, 65, 67,
    74, 77–101, 105–134, 138–140, 143,
    149–152, 156, 159
Luxury features, 15, 44–48
Luxury industry, 7, 23, 28, 33, 35, 139
Luxury items, 1–3, 5, 30, 33–35, 48, 51–53, 65,
    66, 86, 101, 124, 159
Luxury markets, 11, 27, 30–35, 65, 74, 124
Luxury re-positioning, 70–74
Luxury shoppers, 1–24, 27–29, 32–35, 39, 41,
    44, 46, 48, 52, 57, 58, 66, 71–75, 77, 79,
    84, 93, 95, 105–107, 110, 112, 117, 118,
    125–134, 143, 147, 156, 159

M
Ma Nuo, 6
Magnetic sense, 48, 77, 81–91, 101
Maison de Couture, 137–139, 146, 147
Male-to-male competition, 1, 5–7, 16, 23, 31, 41
Marie Claire, 52
Mate copying, 79
Mate selection, 79–81
Maxim's de Paris, 147–148
Melanopsin receptors, 59–60, 75
Memory profiles, 115–117, 125
Michelin star, 99, 111, 117–118, 144
Milan, 62, 123, 128–130
Miley Cyrus, 150
Millionaires, 2–5, 65
Moncler Gear, 41–43
Montblanc, 6, 99, 132, 138, 139, 149–151
Motion, 20, 21, 28, 35–44, 48, 112, 115
    profiles, 37–40, 44
    receptors, 35–37
Moutai, 4, 28, 35, 43
Must-have, 31, 45–48, 51–53, 60, 70, 95, 146, 159

N
Nespresso, 51, 66–69, 74, 133
Neuro-muscular control, 37, 39
Neurosciences, 30–35, 83
Nicki Minaj, 54

O
Opinion leader, 15, 75, 110, 133, 137, 139, 147,
    149–154, 156
Ornaments, 5–6, 16, 17, 24, 55–56, 66, 71, 80,
    82, 86, 101, 107, 141

P
Pan Hong, 149, 152–154
Persona framework, 1, 2, 17, 21, 22, 24
Pharrell Williams, 99, 146
Philippe Starck, 40
Photoreceptors, 83, 89–92
Physiological, 27, 30–31, 33, 40, 48, 56, 62, 71,
    73–75, 110, 139, 140, 143, 159
Physiology, 29, 54–56, 60, 61, 74, 79, 108, 112
Pockets, 9, 52–53, 70
Polarizing, 77, 82–95
Polarotactic Profiles, 86–89, 95
Polarotacticity, 89–91
Porsche, 51–53, 66, 67
Positioning, 18, 51–75, 83, 117, 125, 159
Positioning map framework, 67
Profitable customers, 17–18
Profitable markets, 27–48

R
Raffles Dubai, 130, 154–156
Ranking, 6, 16, 43, 44, 55–56, 107
Relational aggression, 51, 54
Remo Ruffini, 41, 42
Richemont, 14, 56, 99, 105, 125, 132, 133, 150, 151
Rihanna, 146
Rivalry, 6–7, 11
Roberto Cavalli, 105, 128–130
Rods, 36, 40, 57–60, 72, 75, 90, 142
Runway, 52, 153–154

S
Salman Khan, 12, 16
Sense of colors, 56–62
Sense of motion, 20, 21, 28, 35–41, 43, 44, 48
Sense of smell, 137, 141–145, 156
Sense of sound, 1, 10, 15
Sense of time, 106, 111–117, 134
Sense of touch, 1, 9, 10, 37, 74
Sense of vibration, 1, 7–11, 16
Sex ratio, 6–7, 31
Shah Rukh Khan, 32, 141, 149
Shanghai Tang, 16, 57, 68, 99, 106, 117, 124,
    132–133, 150
Sheikh, 28–31, 33, 34, 44, 46, 48, 129, 140
Sheikh Mohammed, 30, 31, 33, 34, 140
Shiny, 6, 52, 57, 74, 77, 79–93, 95, 111
Shoes, 18, 34, 38, 39, 45, 53, 54, 63, 64, 66, 81,
    88, 89, 96, 101, 110, 120–123, 148, 150,
    155, 159
Signature sound, 11–12

Skin spectrophotometry, 80
Smell, 15, 21, 80, 137, 139, 141–148, 156
Social media, 9, 17, 33, 34, 51, 63, 66, 74, 86,
    123, 124, 139, 149, 159
Sofitel, 31, 78, 84, 96–99, 161
Sound, 1, 6–17, 24, 39, 59, 112–114, 132
SpaceX, 13
Status-seeking, 3, 27, 29–32, 35, 43, 44, 48, 148
Stella McCartney, 28, 34, 44–45, 56, 63, 89
Sub-genders, 107–110
Superyacht, 27–29, 31, 39
Swarovski, 16, 44, 67, 68, 78, 89, 93, 99–100,
    138, 150, 154

T
Tabatha Coffey, 137
Taylor Swift, 33, 34, 54, 137, 141, 146–147
Territory, 80, 105, 107–111, 115, 123, 125,
    130, 134
Tesla Motors, 15
Testosterone, 1, 5–6, 55, 71, 73, 81, 108, 109,
    112, 140, 159
The Bear, the Monkey, and the Banana,
    115–117
Tiffany, 16, 68, 72, 77–79, 89, 93, 106
Time, 7, 8, 12–14, 18, 28, 32, 33, 38, 40, 41,
    45–47, 52, 54, 59, 60, 62, 63, 67, 70, 73,
    80, 82, 83, 87, 92, 95, 96, 98, 99, 101,
    105–108, 111–119, 125, 126, 128, 131,
    134, 138–141, 143, 146, 150, 154, 159
    modifiers, 111, 113–115
    and place neurons, 112–113, 126
Touch, 1, 7–11, 17, 24, 37, 43, 59, 72, 74, 97,
    121, 123, 132

Travel retail, 43, 105, 106, 132–134
Trends, 3, 15, 32, 34, 53, 55, 70, 71, 74, 93, 96,
    99, 137, 139, 153–156
Tuhao, 2, 3, 5, 6, 8, 22, 63

U
Ultra high net worth individuals (UHNWIs), 27,
    29, 40
Ultradian rhythm, 111–112

V
Vibration, 1, 7–11, 13–16, 24, 37, 39, 59
Vibrator profiles, 9–11, 16–17
VIP room, 78
Virtual Reality (VR), 40, 44
Vision, 21, 40, 56, 57, 59–62, 66, 67, 73–75,
    78, 84, 91, 96, 118, 121, 123, 133,
    154, 156
Visit Dubai, 149–150

W
Wait Marketing, 46, 105, 106, 118, 119, 121,
    125–134, 139, 151, 159, 161, 163, 165
Wait Marketing 6Ms, 130–132, 159
Wang Jianlin, 29
Wealthy, 5, 6, 30, 33, 40, 52, 53, 71, 79–81, 139
White powder, 40–41

Y
Yachting, 27, 29, 40, 46, 148
Yohji Yamamoto (Y-3), 51, 62–63, 89, 95